ALSO BY KŌHEI SAITŌ

Slow Down: The Degrowth Manifesto

CAPITAL FROM ZERO

Reading Marx in the Age of Climate Catastrophe

Kōhei Saitō

Translated by Brian Bergstrom

ASTRA HOUSE

Astra House

A Division of Astra Publishing House

astrahouse.com

Printed in the United States of America

Library of Congress Cataloging-in-Publication Data is available upon request.

ISBN: 978-1-6626-0341-9

First edition

10 9 8 7 6 5 4 3 2 1

Design by Alissa Theodor

The text is set in Adobe Caslon Pro.

The titles are set in Futura Now Text.

CONTENTS

CAPITAL FROM ZERO

PREFACE

Capital and Red Ink

Why, exactly, did you pick up this introduction to Marx's *Capital*?

Perhaps you feel detached from your work, or are barely living paycheck to paycheck. Across the United States and Europe, the cost of housing has become the central pressure point of everyday life. Rents jump at renewal. Mortgages tighten as interest rates rise. In many cities, the question is no longer how to "get ahead," but how to avoid being priced out—of a neighborhood, a commute, even a basic sense of stability.

Even those with full-time salaried jobs can feel they are barely making it. With both partners exhausted from work and domestic duties, there's little time left for family or rest. Efforts to cut back on overtime run up against the household budget: rent, childcare, student loans, medical bills, and energy costs. For many, the month is a calculation of what can be postponed, what can be paid, what must be endured. The promise that hard work will lead to security rings hollow when insecurity has been built into the system itself.

The COVID-19 pandemic exposed the true extent of inequality in contemporary society. In the same weeks that millions lost hours, jobs, and health coverage, wealth surged at the top. Luxury goods sold out while food banks expanded. Stock markets recovered faster than people did. Entire sectors were declared "essential," yet the people keeping them running were treated as disposable.

The environment is suffering, too. Heat waves, floods, droughts, and wildfires are intensifying across continents. As climate disruptions compound existing inequalities, millions are driven from their homes, food and water shortages become more frequent, and the burden of disaster is borne most heavily by those with the least power and the fewest resources.

It's hardly unjustified to feel upset and overwhelmed about the state of the world. Indeed, what we're witnessing now is the continued betrayal of the promise of increasing wealth and prosperity as the twenty-first century wears on. As people become increasingly frustrated with widespread inequality and the looming effects of climate change, calls to "reform" capitalism and warnings about its limitations and dangers have only become louder.

At this point, I'd wager that most people who've picked up this book generally agree that humanity is standing at a crossroads. But when told to turn to Marx's work to better understand and possibly find solutions for the current crisis, many become skeptical. *Marx? Now? And not just Marx, but communism—throwing away the wealth and freedom brought by capitalism? Are you crazy?*

I fully understand people's skepticism regarding Marx. The first volume of *Capital* was published in 1867. This was only two years after the end of the American Civil War and was a pivotal year in the Reconstruction era. There was no internet and no money markets (at least in the modern sense of the term). Anyone would hesitate, thinking, "Why should I make myself go to the trouble of reading some book about capitalism from a hundred and fifty years ago—what could it possibly offer me now?"

Many people think it's futile to try to understand Marx and argue that the fall of the Soviet Union is indisputable historical proof of Marxism's failure. And it's true that the fall of the Soviet Union, which so extolled the virtues of Marxism and socialism, ultimately weakened leftist movements all over the world.

This was a failure of the grand narrative offered by Marxism, one that attracted so many to its cause: a socialist revolution that would liberate workers everywhere. In its place, capitalism has continued to develop at a rapid rate, promising to liberate humankind from the demands of labor via AI and robotics, to make space travel possible, and to allow us to live longer than ever before, thanks to advances in genetic engineering. The capitalist class also assured us that the planet's environment was going to be managed sustainably amid all this growth.

But how have things actually turned out? Critics of capitalism disappeared, allowing a form of free market fundamentalism known as neoliberalism to reign supreme, which resulted not only in the dramatic widening of the gap between the haves and have-nots, but also the ruination of the planet at the hands of global capitalism. The "dream technologies" like AI and automation promised by capitalism to eliminate the need for labor have yet to produce significant results, with little sign they ever will.

Yet there are still few historians, scientists, journalists, and world leaders in Japan, the United States, and the rest of the world willing to criticize capitalism explicitly and propose ways of overcoming it. Why is that?

As a way to think about how we've reached this moment, let's look at a famous joke: There once was a man from East Germany who was exiled to Siberia. He knew that any letter he sent to his friends would be read by a censor. So he told his friend, "I'm going to use the following code in my letters. If my letter is written in blue ink, that means its contents are the truth. But anything in red ink is false." A month later, the friend received a letter and saw that it was entirely written in blue ink. It said, "My life here is wonderful. The food is plentiful and delicious. The theaters are full of interesting Western films. My living quarters are spacious and beautiful. There's only one thing I can't buy: red ink."

This is a joke about Soviet-style "socialism," but doesn't it apply equally to our lives under capitalism? Aren't the "wealth" and "freedom" we enjoy just like the world described in the letter written in blue ink? For example, those of us living in the Global North can order delicious food via Uber Eats, watch whatever films we want on Netflix, buy a Roomba and let it clean our apartment for us. It might very well appear that we've been given all the freedom we could ever want.

But what if we think of it as the world described by someone deprived of red ink, and thus unable to identify and critique the world's injustices? Because the fact is, incomes are low and the work is tedious. Houses and cars are only available for purchase by those who take out loans—and for many, not even then. You might very well spend your whole life working as hard as you can and still end up deprived of a pension. And on top of all this, the climate crisis is worsening, threatening the lives of billions across the globe. The threat of inflation renders the future of the world economy perilous, while in Japan, there are specific issues like the devaluation of the yen and a dwindling population. The indefinite continuation of capitalism surely fills fewer and fewer people with hope for the future.

In principle, nothing particularly terrible should happen to you if you find yourself fed up with the reality of capitalism and skip work to go for a drink in the middle of the day. But in our current context, to continue to work tirelessly under the assumption that the economic growth and technological advancements brought by capitalism will save us all from climate change and poverty—is this not simply an effect of red ink being unavailable under capitalism as well, leaving us in a blue-ink-covered world?

What we need now is that very red ink. And that red ink can be found in this old text we so blithely threw away—Marx's *Capital*. Why? Because reading *Capital* gives a precise name to the challenge we're currently facing. And thus it represents a necessary first step toward regaining our lost freedom.

Now, I must admit, at the risk of taking the wind out of our sails, that reading *Capital* is no easy thing. It's a thick tome, and written in a very particular, difficult style. There are surely many who will find themselves struggling with the philosophical, political, and economic terms used throughout. But don't worry—the book you're holding is meant to be an introduction assuming no prior knowledge, a reading of *Capital* starting truly from zero.

At the same time, it's a reading of *Capital* "from zero" in another sense as well, a reappraisal from an entirely new viewpoint based on the latest research into Marx's thought, an invitation to imagine together how Marxism can help us forge a new path for the future and build a society beyond capitalism.

1

MANIPULATED BY COMMODITIES

THE STORY OF THE SOUTH SEAS FISHERMAN

Have you ever heard the story of the diligent urban worker who meets the South Seas fisherman?

> "Why do you spend your days lazing around like that?" asks the city dweller. "You need to pull yourself together, work harder, and make more money!"
>
> "Let's say I did—what would I do with all that money?"
>
> "If you save up enough, you can retire and spend all day fishing!"
>
> "But that's what I do already . . ."

Why do we spend our days working so hard, making ourselves so miserable? So we can eat delicious food and laze around on the beach? So we can spend our golden years free of financial difficulty? If so, it really does appear as though the South Seas fisherman lives a much fuller and richer life than we do.

There are, of course, some who enjoy their work and are satisfied with the conditions and the amount of time they have off, including those who work part-time or in temporary positions. But there are many more who are unfulfilled by their jobs and dissatisfied with their lives. It is in this tension that the problem of class emerges.

But before highlighting class differences, it's important to draw attention to the commonalities that unite us. All of us, fisherman or salaryman, have to work to live.

This is why I want to start by examining the meaning of what we call "labor" and begin to uncover the inner workings of capitalism.

LABOR AS METABOLISM

Like all living things, humans make their lives by acting upon nature to create shelter, clothing, food, and other necessities, transforming it in the process so that we can satisfy our needs and desires. Marx referred to this relationship as "the metabolic interaction between nature and man."

"Metabolism" is a term that originated in chemistry and biology and is defined as the process by which something is incorporated into an organism, undergoes various chemical changes, and then leaves that organism transformed into something else. This should be familiar to all of us: It's the process by which food enters the body and is digested, some of it becoming nutrients used by the body while unnecessary parts leave the body as waste.

The concept was first properly defined in 1850 by the German chemist Justus von Liebig (more on him in chapter 4). Marx was an avid student of von Liebig's work, and he incorporated metabolic thinking into *Capital.* At the time of Marx's writing, the concept had already escaped the confines of the natural sciences and was in wide use in fields like economics and philosophy.

"Metabolism" is integral to understanding the fundamental message of Marx's *Capital.* Much of this book's approach to interpreting *Capital* is based on an analysis of Marx's metabolic thinking. In the section "The Labor Process" in *Capital*'s first volume, Marx defines the term as follows:

> Labor is, first of all, a process between man and nature, a process by which man, through his own actions, mediates, regulates, and controls the metabolism between himself and nature. (283)[1]

If we think of labor simply as the exploitation of workers, we might miss this crucial passage. We will discuss exploitation later, but here Marx is speaking more generally, pointing out the inextricability of labor from the metabolic processes of the world.

Those of us living in urban environments tend to forget this, but even things like computers and instant ramen cannot be produced without manipulating the natural world. As Marx puts it, the metabolic relation to nature is "the universal condition" for human existence (290). We can never fully remove ourselves from this metabolic exchange, and this is why labor, too, will never disappear.

WHAT'S SPECIAL ABOUT HUMAN LABOR?

All organisms, not just humans, exist within a metabolic interaction with nature. Humans may cut down forests to make their homes, but tropical termites make homes of their own, intricate hives able to keep their internal temperature steady in astonishingly complex ways.

But Marx points out that there is a decisive difference between humans and other organisms in this regard. Among the organisms populating the planet, only humans participate in the metabolic interaction with nature by wielding their labor as a conscious activity with a clear purpose.

Humans do not relate to the natural world purely according to instinct. Plenty of nonhuman animals devise ways to satisfy their instinctive desires. Leaf-cutter ants in South America, true to their name, cut up leaves, bring them back to their nests, and then use them to cultivate fungus to survive on. Chimpanzees use tools to hunt, forage, and defend themselves, not dissimilarly from humans.

But humans don't, for example, make clothing with the sole purpose of protecting themselves from the cold. They dye the cloth as well to make the clothing more beautiful. They make pottery not simply to facilitate cooking and eating, but also to satisfy needs that clearly lie outside the purview of these instinctive urges; for example, they create earthenware dolls and statues to satisfy ritual or artistic needs.

It's clear from the historical development of technology and culture that unlike all other organisms, humans have multifarious and dynamic ways of "acting on nature."

There is one more crucial thing to remember about this process: The metabolic relation between humans and nature is a cycle—humans' interaction with nature is never the end of the story. When we discard huge amounts of garbage into the environment, for example, it will eventually return to us in some form. Just look at the microplastics coming back to us lodged in the flesh of the fish on our dinner tables. Or consider how the massive consumption of fossil fuels has released huge amounts of carbon dioxide into the air, resulting in the climate change that now threatens our very civilization.

Put another way, the very shape of our lives, of our society, as well as that of the natural world that surrounds us all, is determined by how we act upon nature. The more we mismanage that relation, the more we put both nature and society writ large in jeopardy. This is why labor is essential to humankind's continued freedom and prosperity.

Marx focused so intently on the concept of labor not only because he wanted to expose the exploitation of the working class, but also because he wanted to thoroughly examine this concept of metabolism and work to define it as the *fundamental relationship* between humans and nature.

If we allow ourselves to be unspecific about our understanding of this concept, we risk underrecognizing its urgency—every period in history has involved the management of this metabolic exchange between humans and nature, after all. But the *way* this management takes place differs greatly depending on time and place.

This book will explore how exactly Marx analyzed labor as a conscious, deliberate activity that is managed under capitalism, underscoring the ways in which the relationship between humans and nature allowed Marx to criticize the distinctive nature of capitalism itself.

CAPITAL BEGINS WITH WEALTH

With this in mind, it's finally time to open up *Capital* and begin reading in earnest.

> The wealth of societies in which the capitalist mode of production prevails appears as an "immense collection of commodities"; the individual commodity appears as

> its elementary form. Our investigation therefore begins with the analysis of the commodity. (125)

These are the opening lines of the first section of the first chapter of *Capital.*

While this opening section may be titled "The Commodity," by looking closely at these opening lines we can see that the subject of the first sentence is not "commodity" but "wealth." In other words, while the *analysis* contained within *Capital* will begin with the commodity, *Capital* itself begins with wealth.

Marx is saying here that the wealth of "societies in which the capitalist mode of production prevails"—that is, capitalist societies—*appears* in the form of commodities. As we saw earlier, Marx attempted to illuminate the distinctiveness of capitalism, and how capitalist societies differ from other types of societies, by examining labor as its foundation.

In capitalist societies, labor produces commodities. And indeed, the world is overflowing with all the commodities we produce. But in noncapitalist societies, the wealth produced by labor doesn't necessarily appear in the form of commodities. Therefore, to understand capitalism, we must analyze commodities and the labor that produces them.

WHAT IS WEALTH?

Let's first consider what "wealth" really is. What likely first comes to mind is money, stocks and bonds, real estate, and the like; it's a word that conjures images of riches whose worth is measured in and expressed as money. But such a definition blurs the distinction between wealth and commodities. And indeed, we live in a world that has instilled this simplistic equivalence as a type of common sense.

As the subtitle to *Capital* makes clear, one of the chief aims of Marx's project is to untangle these sorts of ideas, expanding habitual forms of thought and narrow worldviews as part of his "critique of political economy."

So let's try a different approach to the question of what we really mean by "wealth." The German word for "wealth" is *Reichtum*. *Reich* is cognate with the English word "rich," and can of course mean "rich" as in a "rich man," someone who possesses the type of wealth measured by money. But it can also refer to things like smells and tastes; richness in this sense means something closer to "ample" or "plentiful"—that is, *abundant*. In other words, something that cannot be measured by money.

Natural features like air or water can be *abundant*, and this richness can be thought of as part of a society's *wealth*. Beautiful green forests, parks where anyone can go and spend their free time, local meeting halls and public libraries—things like these are important assets for society, representing important sources of "wealth." Knowledge, art, culture, communication skills, practical workplace skills: All of these are forms of wealth as well. We can't always measure their worth with money, but these things are necessary to live a rich, rewarding life, and thus can and should be considered essential parts of a society's wealth.

THE COMMODITY'S TRUE FORM

Labor, then, is what creates, sustains, and develops wealth. (Repair and maintenance are also integral functions performed by labor.) Humans cooperate with one another, acting on nature to advance our skills and transform the natural world according to our desires, enriching our stores of wealth. Marx reminds us, though, that at some point this type of societal wealth became transformed, with the rise of capitalism, into commodities.

We can see this happening all around us still. For example, city parks are renovated to become high rise condominiums and malls, and forests are cleared to make way for golf courses or mega-solar farms.

An easy way to understand this process is water. When I was a child, drinking water was not a commodity, but was instead available for free from the tap. About thirty years ago, though, water became something to be put in plastic bottles and made into the commodity known as "mineral water."

Subsequently, some people began avoiding tap water even though its quality hadn't changed. Water that was once enjoyed for free without a second thought became something we paid money to drink. Our feelings and desires were transformed by the commodification of society, resulting in increased profits for industry and increased burdens on our personal budgets.

One of the most defining features of capitalism is exactly the transformation of anything and everything into a commodity. By now we're so used to this progression that it seems completely unremarkable. But as Marx reminds us, here at the beginning of *Capital*, there is nothing natural about this process, and it is not a precondition for the production of wealth.

As the water example shows us, the emergence of societies within which wealth appears as an "immense collection of commodities" is a relatively recent thing. Even in major American cities, it was normal before World War II for households to grow vegetables in their own gardens, taking what they needed for themselves and sharing the rest with their neighbors. Capitalist societies may seem to have existed since time immemorial, but a glance at history shows that this is not at all the case.

Of course, commodities existed even in precapitalist societies. But most were luxury goods or goods made for trade, while everyday necessities were, as a rule, made and gathered within communities to be shared among everyone. Goods that one had to pay money for were limited to certain domains; societal wealth was far from being defined as an "immense collection of commodities."

But what about now? We live in a world in which most of the things we need to survive are commodities, and it's almost entirely impossible to navigate the world without encountering commodities in some capacity.

Our world is overflowing with a diverse array of attractive commodities. We can spend money and have any of these marvels at any time. Our lives may feel like they have become "rich" as we accumulate commodities, but, as Marx argues, this kind of commodification comes at the explicit expense of social prosperity.

THE ENCLOSURE OF THE FOREST BY CAPITAL

After completing his doctorate at the University of Jena in Germany, Marx began working as an editor at the newspaper *Rheinische Zeitung*.[2] In 1842, he became editor in chief and began publishing extensively about "wood theft."

In Germany at the time, poor farmers would customarily gather fallen branches from nearby forestland to use for cooking and to heat their homes in winter. These fallen branches were an indispensable form of "wealth" for them. A law was made, though, that redefined this activity as "theft," causing wood gatherers to be driven off the land by police on horseback.

The 2017 film *The Young Karl Marx* memorably reenacts this incident in its opening scene, showing how even fallen branches had become a form of private property owned by landlords, and illustrating the merciless violence inflicted on poor farmers that was justified by the proclamation "If you want these branches, you have to pay for them!" Until that point, anyone could freely gather whatever branches they needed without worry, but now even this marginal activity was infiltrated by privatization and commodification. Marx was shocked. He fiercely criticized how dramatically the logic of the commodity had dominated German society, eventually drawing the attention of the authorities, which ultimately led to the dissolution of *Rheinische Zeitung*. As a result, Marx and his family left Germany and moved to Paris.

The criminalization of the German farmers illustrates how the transformation of a society's wealth into commodities is, simply put, the process by which a price tag is affixed to things, marking them as "for sale."

The wealth held in common by everyone, that everyone can access for free, is thus monopolized by capital and turned into the object of monetary trade; in other words, it becomes a commodity. For example, a bottled water company will buy land that has freshwater springs that they can then bottle and sell as a commodity. The water sources once used by local people thus become inaccessible, forcing them to go to the supermarket and buy that same water.

Those with money can, of course, buy whatever they want, but those without money often can't afford the bare essentials. Think of health care in the United States. In a country that boasts the world's largest economy, there are naturally enough doctors, hospitals, and medication for the entire population. Yet the cost of medical care is so high that those without sufficient means often can't go to the hospital when they need to. Thus, health care has been made artificially "scarce" and businesses like health insurance providers and pharmaceutical companies reap huge profits.

But why on earth would capitalism operate in such an absurd, irrational way?

THE TRANSFORMATION OF THE RELATIONSHIP BETWEEN HUMANS AND NATURE

Former commons like forestland and water were abundant forms of wealth as anyone could access them. It's this quality that makes the commons anathema to capitalism. If people can acquire whatever they need to live without money, then no matter what commodities one might make, they ultimately won't sell. Thus, under capitalism, the commons must be monopolized or even partly destroyed so they can be made into things that are then bought and sold; that is, they must be made into commodities.

Furthermore, even if you shut people out of a forest to monopolize it, you still have to cut the trees down and transform the wood to commodify it. In other words, commodities necessitate labor. And who will perform that labor? The very people you drove out of the forest, who now need money to buy the wood they used to access for free. In this sense, disassembling the commons benefits capital in two ways.

Landowners and feudal lords in fifteenth- through seventeenth-century England illegally partitioned farmland, driving the farmers off this stolen land so that they could turn it into pastures for sheep, which at the time were more profitable than crops. This is what's known as "enclosure." Many of the farmers driven off their land became vagrants who traveled the countryside. These dispossessed and impoverished farmers led revolts against the landowners all over the country.

In the England of Marx's era as well, the landowning class was partitioning land again, effecting a "second enclosure" that was legal this time, supported by parliamentary legislation. People were driven again from their farmland, and, deprived of anywhere to live or any way to get by, they moved to cities in search of work, becoming the backbone of the population of factory workers. This was the formation of the proletariat.

But this is not to say that the displaced farmworkers instantly transformed into docile wage workers. They first became vagrants and criminals, upsetting law and order within the cities. Government and police forces intervened, arresting the transients and bandits and putting them into workhouses to use as forced labor. This violent historical process ultimately gave way to the docile working proletariat.

On the other hand, the farmers who managed to stay and continue cultivating the land became wage workers as well, employed by the large-scale agriculture industry to produce crops as saleable commodities. In this way, the marketplace expanded steadily to take over all elements of society.

> In fact, the events that transformed small peasants into wage-laborers, and their means of subsistence and of labor into material elements of capital, created, at the same time, a home market for capital (910).

Those who produced commodities not only gave up their labor power, but also became consumers of commodities as well. This is how the market economy began to emerge, allowing landowners and capitalists to profit more than ever and thus ushering in the development of capitalism itself. This was the formation of the bourgeoisie.

Marx called this violent process "primitive accumulation," describing it as a unique process of theft under capitalism that dismantles the commons and drives people into wage labor. This is also the moment when humans become separated from earth's natural wealth and their lives become impoverished and unstable.

Recall the discussion of metabolism that opened this chapter. As capitalism develops and wealth is transformed into commodities, the metabolic exchange between humans and nature takes on a completely different shape than ever before.

THE INABILITY TO STOP SEEKING PROFIT

Let's look more closely at the differences between capitalist and noncapitalist societies.

In capitalist societies, every form of wealth is transformed into commodities. But as we saw when we analyzed the opening lines of *Capital*, this process remains at the level of appearances. We must take our thoughts a step further now and shift our analysis to investigate that which produces these commodities—namely, labor.

Humans have always created things using labor, but according to Marx, before capitalism, labor was fundamentally a way to satisfy human needs and desires. For example, humans cleared land to cultivate grains and vegetables in order to satisfy their hunger. They made clothing that would protect them from the cold and rain. Making beautiful clothing to indulge creative urges, constructing grand temples as a display of political power, starting wars to satisfy a king's desire to expand his realm—these, too, are all specific objects tied to specific wants and needs.

People perform labor (or make others labor for them) to satisfy these concrete desires, but this sort of productive activity has a natural limit. No matter how much you want to eat, there's a limit to how much you actually can. If you produce too much food, it will go bad; even the most ostentatiously power-hungry king won't end up constructing hundreds of grand temples during his reign.

But the capitalist society we live in loves nothing more than money. It can conveniently be exchanged for anything at any time. Money doesn't go bad, and if you buy stocks with it, it will increase by itself. Here we see the birth of the desire for limitless wealth, the primal force behind capitalism's rise.

Because money possesses the special power to "buy anything," those with more of it naturally become more powerful. Those with sufficient assets can make a killing manipulating stocks; you can neutralize the threat of a company potentially becoming a rival by buying it while it's still small. Donate enough money and politicians will listen to you; they may even create advantageous tax laws or implement other beneficial legislation for you.

As a result, one sector of the population ends up monopolizing wealth, creating deep divides within society. According to the international NGO Oxfam, as of 2018 the world's twenty-six richest individuals held as much wealth as the poorest half of humanity—3.8 billion people.

This inequality exists not only among nations, but within them, too. Even if we narrow our focus to only the United States, for example, the wealth possessed by the top fifty richest people is valued at around two trillion dollars, equivalent to that possessed by the poorer half of the country's population: 165 million people. The CEO of Tesla, Elon Musk, is the world's richest man, possessing assets valued at over two hundred billion dollars.

But it's not only Musk. Microsoft founder Bill Gates, Amazon's Jeff Bezos, and Meta (formerly Facebook) CEO Mark Zuckerberg all have more money than they can ever use, yet they still strive to get even richer. No number of private jets or mansions or yachts could exhaust their resources, so now they compete to develop rockets to travel into space.

On the other hand, the average person battles long work hours, unstable employment, and low wages. Even working as hard as they can, many cannot manage to put away any money for their savings or build a family. Because of how costly health care, rent, and education have become, many feel extremely precarious in their lives. Inequality and the high cost of living have divided Americans culturally and politically, giving way to Trumpism and the rising popularity of other extremist ideologies.

How did it come to this? The main reason is that the purpose of a capitalist society isn't the satisfaction of human wants and needs; rather, it's the increase of capital itself. We'll examine the mechanism by which this happens in more detail in the next chapter, but capitalism cannot stop profit seeking. Whether it leads to destructive results like the eradication of bookstores from the public marketplace or constructive results that increase social wealth, the purpose of capitalism is solely the unending pursuit of profit.

NOT THINGS WE NEED, THINGS THAT SELL

If the overriding purpose of production shifts from satisfying human wants and needs to increasing capital, then it follows that both the mode of production and that which is produced will shift as well.

The commodities produced in a capitalist society are judged not by whether they're truly necessary or important to people's lives and happiness, but rather by how many might sell and for how much—that is, how much they will help increase capital.

A prime example of how this works is the way trendy bubble tea shops or fancy bakeries suddenly pop up all over only to disappear again seemingly overnight. Another example might be the shortages of masks and hand sanitizer during the COVID-19 pandemic. Even though experts in infectious disease had urged the production of such goods as precautionary measures before the pandemic, they were deemed "wasteful" as commodities (since they would not immediately increase capital) and were not produced in sufficient quantities to fulfill the public's needs once the crisis occurred.

This is not meant to be an indictment of individual companies involved in this production. Rather, this is the nature of capitalism itself. Industry had no choice but to make things that will sell, not things like masks that have only a limited immediate demand in pre-pandemic times. Once the pandemic hit and everyone needed masks, recall that all sorts of companies—sportswear makers, fashion houses—jumped in to produce them; the mask market went from one plagued by shortages to one that was completely saturated. Now there were too many masks, and they began to be sold at steep discounts just to get rid of them. This is a typical pattern within capitalism.

As we can see, necessary things and profit-generating things are not always the same. Marx speaks to this, pointing out that the commodity has two faces.

One face is called "use-value." Use-value is the usefulness of a product to humanity; in other words, its ability to satisfy human wants and needs. Water possesses the ability to quench thirst, food has the ability to fill an empty stomach. The use-value of a mask lies in its ability to mitigate the spread of contagious disease. The creation of use-value necessary to sustain life was the main purpose of production in precapitalist societies.

For capitalism, however, it's the commodity's other face that is most important: its "value."

To become a commodity, a good must enter the marketplace and be exchanged for money. A chair that has never been bought or sold possesses only its use-value—its ability to be sat on. By contrast, a chair that has become a commodity has a price tag—for example, seventy dollars—that makes it interchangeable with anything bearing the same price: two hundred eggs, say, or a set of bedsheets. How can this be?

The use-values of chairs, eggs, and bedsheets are completely different. Which is more useful, an egg or a chair? If you're hungry, the egg would likely be more helpful to you, while if you need to do some work at your desk, a chair would be. Comparing the two according to their use-value sheds no light on why two hundred eggs and a chair would both be priced at seventy dollars. Neither will it explain why that price is seventy dollars and not, say, thirty-five dollars.

The *value* of a chair, two hundred eggs, and a set of bedsheets is the same, though, expressed as "seventy dollars." According to Marx, this sense of value is determined by the amount of labor time needed to produce something. In other words, it takes the same amount of labor to produce a chair, two hundred eggs, or a set of bedsheets, and that's why they possess the same value and thus can all be had for the same price. This is known as the labor theory of value.

In common parlance, value can refer to the importance something has to someone, for example a chair that your grandfather really loved, which makes it valuable to the family. But Marx does not use the word value in that sense, so keep this in mind as we work our way through *Capital*.

Value and use-value might be easy to confuse as well. But they are completely different—think, for example, of air, which has an extremely high use-value as something necessary for life itself, and yet is free, and then think of a diamond, which has a low use-value but is extremely expensive. Air exists independently of human labor, so it has no value. But mining for diamonds is incredibly labor-intensive, and so their value is quite high.

You can directly experience the use-value of something just by using it, but value is something indetectable by the five senses. Marx calls it a "mystical" characteristic (164). Because commodities appear in our everyday lives with price tags already attached, we might feel like we more or less have a grasp on what they express, but the fact is, these commodities around us possess an invisible, mysterious power. And this mysterious power—namely, value—only becomes stronger as the marketplace expands.

HUMANS USED AND MANIPULATED BY OBJECTS

Once we distinguish between the commodity's two faces, we can start to explain the mechanisms by which capitalism gives rise to various contradictions and irrationalities.

Under capitalism, the most important thing is how much something might sell—how much money might be made from it. Value, appearing in the form of the price, is prioritized above all, while the relevant use-value is only of secondary concern. For example, in consideration of not only the planet but also one's pocketbook, the commodities we create should be durable and should be made from sustainable materials. But look at the phenomenon of fast fashion, which fills our closets with clothes made as cheaply as possible with no regard for environmental costs. Here we see how "things that make money" (value) and "things necessary for living" (use-value) become completely severed from one another.

It's not only consumers who are manipulated by value. Holders of capital are driven to produce large amounts of "things that will surely sell," and if they fail to be in demand as predicted, they're left with stores full of useless stock and wind up bankrupt. Even if a product does become a hit, as in the example of the bubble tea shops, competitors soon pop up all over, overloading the market and leading to plummeting sales and financial difficulty.

Observing this cycle, Marx states:

> The value character of the products of labor becomes firmly established only when they act as magnitudes of value. These magnitudes vary constantly, independently of the will, foreknowledge, and actions of the exchangers. Their own movement within society has for them the form of a movement made by things, and these things, far from being under their control, in fact control them. (167–8)

As this passage indicates, value moves constantly on its own, like the weather, completely independent of our thoughts or feelings. This means that even though people make things to use them, the magic of value transforms these things in unpredictable ways, and people ultimately become manipulated by commodities.

When things were made for their use-value, people of course used them, just as the term implies, but under capitalism, the drive to make things in order to create value reverses these positions: People end up used by things. This is the process Marx calls "reification." As soon as an object produced by human labor becomes a commodity, it gains the mysterious power to take over and control human life and activity.

Even though we all collectively depend on the things we create, no one coordinates production across society as a whole. Rather, everyone labors as individuals, which results in us frequently producing a surplus of things no one needs or too little of things everyone does.

In the end, we bring what we produce to the marketplace and then must decide how much of it to keep producing only belatedly, after gauging how much others seem to like it. This means that the productive activity of humankind ends up manipulated by the vagaries of value.

A DANCING TABLE?

Marx writes the following passage about the uncanny nature of reification:

> The form of wood, for instance, is altered if a table is made out of it. Nevertheless the table continues to be wood, an ordinary, sensuous thing. But as soon as it emerges as a commodity, it changes into a thing which transcends sensuousness. It not only stands with its feet on the ground, but, in relation to all other commodities, it stands on its head, and evolves out of its wooden head grotesque ideas, far more wonderful than if it were to begin dancing of its own free will. (163–4)

Marx's rhetoric might be idiosyncratic, but what he's saying is simple. Let's say you make a table out of wood. Your labor has transformed the wood into the form of a table, but the wood, both before and after becoming a table, remains "sensuously" consistent: It looks, feels, and smells like wood. You have transformed the wood into a table, but as long as all you do is use it—eat your meals on it, sit at it to read a book, place a vase of flowers on it—then it remains simply a table, possessing nothing but use-value.

As soon as you decide to sell it to someone else as a commodity, though, the table "transcends sensuousness"—it becomes something possessing value, a quality indetectable to the human senses. As a commodity, the table enters the marketplace and begins negotiating its place among all other commodities, and in this way coerces people to join in the dance, driving them mad with its movement.

As long as we live in a capitalist society, we have no choice but to let the table take the lead in this strange dance. If the table doesn't sell, it can lead to terrible outcomes for its maker: insolvency, bankruptcy, unemployment, even suicide. On the other hand, if the table becomes a hit, it might make you rich.

People devote themselves, heart and soul, to making any commodities that seem like they will sell. Something that will be used once then discarded, that doesn't really help anyone do much of anything—as long as it sells, people will make more and more of it regardless.

It's easy to see how this works if you go to your typical dollar store. Most of the things sold there are trivial plastic goods.

But once given an appealing name and some cute packaging, they begin to take on their own life and charm. This is how we end up with so many commodities everywhere that have little to no use-value and end up in our landfills as garbage.

As this process continues, so does the rise of advertising and marketing as industries. Consumers aren't total fools; soon enough, they realize they've essentially bought a piece of junk. Once they do, the spell of the commodity is broken, and they tire of the object they purchased. This creates marketers' need to do everything possible to transform a different piece of junk—and then another one, and another—into a brand-new, appealing commodity they can sell.

Let's look at another example symbolizing how reified society has become: the push by President Trump and other largely Republican lawmakers to reopen schools and businesses during the COVID-19 pandemic.

Because of the way contemporary society is structured, a halt in commercial production and consumption could, depending on the case, result in effects worse than the pandemic itself. But this campaign, which actively supported risky activities like airplane travel and eating out, all in the name of keeping the economy "moving," became the object of harsh criticism. To prioritize "moving" the economy over protecting human lives—wouldn't it be more accurate to think of such a valiant effort as one in which the economy was "moving" us, not the other way around?

In Japan, there was also the matter of the Tokyo Olympics, a huge event riddled with suspected corruption that was forced on the Japanese public in the midst of the pandemic to the tune of over $19 billion. In Osaka, the 2025 World Expo brought the construction of new resorts along the coastline and new casinos for people to spend their money in. The new Maglev Shinkansen rail line connecting Tokyo to Nagoya and Osaka, known as "the Linear," has also come under fire for flattening mountains and damaging water sources during its construction.

Meanwhile in the United States and elsewhere, popular digitalization innovations like AI and Bitcoin, which many view as crucial developments for productivity, efficiency, and discovery, pose a serious risk to the environment and our natural resources as they require huge quantities of water, land, and energy to keep the servers running.

Is all this development really making our society richer? It seems as though the main beneficiaries of this boom are consultants, contractors, and hotel developers, while the meager number of low-paying jobs these projects bring to locals amounts to nearly nothing.

Economies are not kept in motion by people; rather, the economy's movement is an end unto itself, and humans are living their lives as cogs in the self-propelling machinery of the capitalist economy. Marx identifies this as the problem at the very heart of reification.

THE ENCLOSURE KNOWN AS PRIVATIZATION

The power of reification gets stronger as the space taken up by commodities expands, so that people are manipulated by things more and more. Its force has become all the more palpable over the course of the past few decades.

The second half of the twentieth century saw the end of high-speed economic growth, and as the capitalist economy stagnated, the privatization of public works and the deregulation of industry proceeded apace, all in the name of free markets. These are the policies known as "neoliberalism." The UK's Margaret Thatcher, for example, believed that anything and everything should be subject to market forces and that competition would lead to overall improvement in all sectors. Yet, as neoliberalism has spread throughout the world, the power of reification has only grown stronger, leading to an increase in monopolies and other unforeseen, irrational, and inefficient results.

Many still imagine that "dictatorial" control by governmental and public institutions is inherently inefficient, and that transferring this control to private hands "democratizes" management, making things more efficient. But this is an illusion created by language. Privatization is the transfer of something publicly owned into private hands. But the root meaning of "private" is "privation" or "withdrawal." It's the withdrawal of the "commons" held by all into the hands of the few. The true nature of privatization is the monopolization of power by specific industries—a modern update of enclosure, meant to expand the reach of commodification even more.

The market is never "democratic." As we have seen, the market is only accessible to those with money. The advance of privatization means that those who were once able to access public services like health care and education are now shut out of them.

Further, because the market privileges profit above all else, goods and services that don't generate enough money end up mercilessly cut, often alongside budgets and staff as well. That which is truly wasteful should of course be subject to cuts, but cuts that don't consider use-value often lead to the elimination or degradation of necessary goods and services, which diminishes the overall wealth of society. An illustrative example of the way overzealous pursuit of so-called efficiency jeopardizes sources of shared affluence is the Japanese public library system.

SOCIETY'S WEALTH IS IN DANGER!

The number of temporary workers staffing public libraries all over Japan has risen precipitously. Over half the staff of the libraries in all prefectures are now temporary workers. In Nagano prefecture, the proportion has reached nearly 80 percent. This sharp increase in temporary employment is a direct result of the pursuit of "efficiency" and "cost cutting" by library management. Libraries will naturally never make money. So the only way to render them more "efficient" is to cut workers' pay, which shifts the burden onto the workers themselves. Recently, a heartbreaking appeal from a temporary staff member in her twenties working at a regional public library to raise her salary to a mere 98,000 yen [approximately 623 USD] per month made waves on social media.

People have continued to work as temporary staff at these libraries for such low wages because they love books and value performing their jobs to the best of their ability out of a sense of mission. But is it right to exploit their commitment and generosity in this way?

If things continue like this, fewer people will want to become librarians, and eventually the quality of the services the libraries offer will decline. In the face of budget cuts and labor shortages, precious archives may go improperly managed. The archives held by a public library are an important type of commons, and a prime example of societal wealth. Knowledge and culture are things that get passed along and developed precisely through universal access. But archives are not commodities; they don't produce profit. Managing a library within the logic of capitalism, which prizes value over use-value, only leads to "reforms" that diminish society's wealth.

There are many other examples of moves toward commodification through the sacrifice of societal wealth. For example, the redevelopment of the wooded area around the Meiji Shrine, known as the Jingū Gaien, became a major controversy after regulations were relaxed to facilitate the construction of the Japan National Stadium. The grounds of the shrine contain stands of trees that are over one hundred years old. The proposed redevelopment would cut down almost a thousand of these trees to build a large new shopping center.

Trees don't make money just standing there growing—why not cut them down and build a multiuse facility where people can shop, eat, and play sports? This sort of enclosure through "redevelopment" will almost certainly generate money.

But is there really such a pressing need to destroy one of the precious few remaining peaceful oases in the concrete jungle of Tokyo? In a society where the population is continuously decreasing, do we really need to build more high-rise condos and shopping centers? In the end, this redevelopment would turn a place where anyone can go to relax and enjoy themselves into a playground for the rich. Following the logic of value so often results in a society becoming a more difficult, miserable place to live.

This is what really happens when the wealth of a society "appears as an 'immense collection of commodities.'"

THE INTERNALIZATION OF COST-PERFORMANCE THINKING

There are surely some readers who, hearing me talk about the Meiji Shrine redevelopment project, think to themselves, "But urban development is normal under capitalism. What's the point in criticizing it? How many trees were cut down over the years to build up Tokyo anyway?" This reflects a major difference in value systems, I would say. What does it stem from?

As we have seen, it's impossible to live in modern society without relying on commodities. We cannot make our lives work without also "moving" the economy. In a society that revolves around money, we assume that our ability to skillfully navigate commodity-based economies constitutes "rationality." In other words, we've internalized the logic of value as embodied by the commodity as our own value system.

As this occurs, the feeling that wealth should be measured by use-value begins to seem hopelessly naive; the highest purpose for production under capitalism is increasing value. Use-value, in this scheme, is dismissed as an ineffective way to create value. Marx warns us that this opposition between use-value and value—that is, the opposition between wealth and commodities—will eventually bring about the demise of both humankind and nature.

We can see how the logic of value becomes internalized in cost-performance thinking. This is a strategy that evaluates efficiency by gauging the "returns" achieved through any investment or activity.

Let's say you're earning an hourly wage of thirty-four dollars. If it takes you an hour to cook for yourself, this means you lose the opportunity to earn another thirty-four dollars. In this sense, the cost of doing something that doesn't earn money, like cooking for yourself, is quite high. Paying $60 to eat out or order something from Uber Eats thus becomes the more optimal choice.

This is how domestic activities like cooking and childcare become commodified. And as more people strive to use their time more efficiently, more service industries pop up to fill this need.

If we follow cost-performance thinking to its logical conclusion, though, all sorts of things begin to seem like an inefficient use of time: neighborhood associations, local festivals, PTAs, even labor unions. Even helping one's family and friends ends up falling into this category. Community begins to wither, and eventually collapse.

I'm not suggesting that older arrangements, wherein domestic duties were left entirely to women and neighborhood associations were forums for elders to throw their weight around, are something we should return to. Rather, I'd like us to remember there are alternatives to both patriarchy and commodification that can enable us to share domestic and childcare duties equally between the sexes and strengthen community bonds through voluntary action.

But instead of choosing these ways of living, our society has instead elected to commodify anything and everything, with every bit of time we have devoted to the pursuit of money.

This paradoxical situation resembles the world portrayed in Michael Ende's novel *Momo*. The barber, Mr. Figaro, decides to save time at the Timesaving Bank and finds himself deprived of any free time at all:

> He was becoming increasingly restless and irritable. The odd thing was that, no matter how much time he saved, he never had any to spare; in some mysterious way, it simply vanished. Imperceptibly at first, but then quite unmistakably, his days grew shorter and shorter. Almost before he knew it, another week had gone by, and another month, and another year, and another and another.[3]

A life devoted to saving time in the pursuit of optimizing cost-performance is not a rich, fulfilling life. Rather, it's one in which free time disappears, along with the ability to communicate and maintain bonds with your family, neighbors, and friends. Society thus becomes a collection of miserable and overworked individuals unable to permit even the slightest "waste" of time.

THE SHORTCUT TO A BETTER SOCIETY

In this atomized society dominated by the individual, there are many who work tirelessly and yet, because they cannot make enough money to get by, they fall into debt, poverty, and unemployment; some are even overworked to death. Can we really write these problems off as a simple matter of "personal responsibility"?

The truth is, much like in the case of the redevelopment of the Jingu Gaien, the intentional deprivation of a universally accessible commons becomes a way to increase the monetary wealth of a select few elites and heads of industry. The continuous expansion of this kind of societal opposition and divide is a defining characteristic of the capitalist mode of production—that is, the production of commodities for the purpose of increasing value and capital. Marx's *Capital* diagnoses this as a structural problem in society, rather than one arising at the level of the individual.

Yet an examination of the process by which money and commodities circulate in the marketplace will not sufficiently illustrate the true scope of the problems lying at the heart of capitalism. We must now shift our gaze to that which gives rise to commodities: production. The following chapter will examine this issue in detail.

Simply understanding the sources of the contradictions within capitalism and the mechanisms by which they function will not, unfortunately, make these contradictions disappear. It will take an enormous change in the way society is arranged to address them—a difficult path still lies before us.

But the role of theory is to prevent us from going down a mistaken path and getting lost. Instead, theory helps us move toward healing our collective ills. Marx explains the purpose behind *Capital* like this:

> Even when a society has begun to track down the natural laws of its movement—and it is the ultimate aim of this work to reveal the economic law of motion of modern society—it can neither leap over

> the natural phases of its development nor remove them by decree. But it can shorten and lessen the birth-pangs. (92)

By revealing the contradictions within capitalist society, Marx sought to provide a "shortcut" to the creation of a better society, one no longer beholden to capitalism.

In the rest of this book, I aim to read *Capital* not as a work of economic theory, but rather as one filled with clues as to how we can solve modern social problems such as social divisions, long work hours, and the like, using Marx's words to illuminate the reality of the world around us.

2

WHY CAN'T WE STOP WORKING OURSELVES TO DEATH?

THE ENDLESS GAME OF VALUE INCREASE

In the previous chapter, we explored the differences between capitalist and noncapitalist societies by thinking about commodities. There are parts of noncapitalist societies that trade in commodities as well, but most of the things they produce are not commodities, but rather items created for their use-value because they help make people's lives livable.

Capitalist societies, by contrast, commodify many different forms of societal wealth—water and forestland, or even community-owned resources like public parks and libraries—transforming them into things that cannot be accessed without money. This means that everyone must devote all their energy into making money in order to survive.

As such, those with money are the ones with the highest status in a market society.

Goods are increasingly produced not for their use-value but because of their value; they are created not because they are necessary but because they might sell.

Under capitalism, there is constant competition in the marketplace to increase value. Capitalism is an endless game of value increase. But does this game create happiness? Not really. The real victims of a capitalist society are workers and the natural environment.

But it's not enough to think about this problem solely by analyzing commodities and money. We must also analyze capital itself.

CAPITAL IS MOVEMENT

What, then, is the nature of capital? Most people hearing the term probably imagine a certain amount of money, or perhaps goods—machinery, buildings, or other assets—worth a certain amount of money.

But according to Marx capital is neither money nor goods—not machinery, not factories, not commodities. It's movement.

What kind of movement? Incessant movement to make itself grow, to increase its own value. Marx expresses this movement in German as G–W–G^1—what he calls the "general formula for capital." G stands for *Geld*, or "money," while W stands for *Ware*, or "commodity." In English, this formula is expressed as M–C–M^1 (money–commodity–$money^1$) (247–57).

In a society where production is based on use-value, a cobbler will make a pair of shoes, sell them, and then use that money to buy bread. He'll eat the bread, and then that's that. Such movement would be expressed as C–M–C.

But in a capitalist society, the cobbler uses his initial capital to make shoes, and then when he sells them, he'll use that money to make more shoes. If they sell too, he'll invest even more of his capital into making more shoes if he thinks they'll keep selling. The difference between M and M^1 is the difference between his initial outlay of capital to make the shoes and the amount of money he makes from selling them, which may include profit.

Capital grows by continuously repeating this M–C–M^1 movement. It doesn't matter whether you make shoes or bags or anything else as long as you're producing profits. Use-value becomes, at best, a secondary concern.

Simply put, capital is movement that makes money, and any society primarily concerned with the upkeep of this moneymaking movement is a capitalist society.

THE REASON WHY CAPITALISTS CAN'T STOP SEEKING PROFIT

To review, the governing principle driving profit seeking is not a commodity's use-value, but rather its value. The movement by which capital grows—expressed as M–C–M^1—increases value's power, enough that value becomes a type of subject in and of itself, able to act on its own and manipulate humans according to its will.

> In the circulation M–C–M both the money and the commodity function only as different modes of existence of value itself . . . It is constantly changing from one form into the other, without becoming lost in the movement; it thus becomes transformed into an automatic subject . . . Value is here the subject of a process in which, while constantly assuming the form in turn of money and commodities, it changes its own magnitude, throws off surplus-value from itself considered as original value, and thus valorizes itself independently. (255)

Money and commodities can thus both be seen as simply temporary forms taken on by capital. Value is generated as capital moves from one form to another; according to Marx, value then becomes the self-propelling subject of this movement.

Let's go back to our cobbler, who's now a proper capitalist. He succeeds in his shoemaking business and makes about $125,000. He now thinks, "Next I'll make $190,000!" If he succeeds in that, then naturally he'll think, "Let's try for $250,000!" Even if he makes $250,000, or more than that—even an amount of money he could never spend over the course of his whole life—he will still end up wanting more. As Marx puts it, "The movement of capital is therefore limitless" (253).

Marx stresses that capitalists are not a special race of people who lust after money, but rather are normal people fulfilling their "destiny" bound to the whims of capitalism. Their behavior isn't a question of morality—those living under capitalism are not free to control their own desires and decide, for example, to use profits to raise the wages of their employees.

Ruthless competition is always taking place in the market. Capitalists toil tirelessly pursuing every dollar of profit they can, constantly working to improve their existing products and bring new ones to the market. Otherwise, they'll stand to lose their share of the market to competitors and wither away, eventually becoming unable to pay their workers at all. Efficiency and cost-cutting are the order of the day.

In other words, even capitalists are nothing more than cogs in the self-propelling machine of capital that strives only to increase its own value. As long as capitalists wish to remain capitalists, they must submit to capital's automatic movement. And by extension, workers have no choice but to submit to capitalists.

The self-propelling movement of capital animates society: Both humans and nature become controlled by it as well, their very existence reduced to being a function of capital. Swallowed up by the limitless movement of capital to grow and valorize itself, everything becomes a cog in the same machine. This creates a severe disturbance in the "metabolic interaction between nature and man."

"THE HIDDEN ABODE OF PRODUCTION"

Before any discussion of how capitalism disempowers and alienates workers and impacts nature and the environment, it is important to understand the mechanism by which capitalists grow their capital and increase its value.

As the general formula M–C–M^1 shows us, keeping your money safely in your pocketbook will never make it grow. It's irrational under capitalism to be a miser.

If you want to be a proper capitalist, you must take risks and invest your money producing things that you believe may sell. This is the only way to make capital grow. The stock and foreign exchange markets may make money for you, too, but as Marx shows, at base, the only real way to increase value is in "the hidden abode of production" (279).

What does this mean? Let's break down the issue to its simplest form.

Let's say a worker works for eight hours and is paid $60 a day. The capitalist pays the worker $60 after receiving a day's work from him. If the value of the commodity produced by that day's work is priced at $100, then the capitalist's profit is the amount left over after subtracting the cost of the labor: $40. This is what Marx calls surplus-value, which is how capital grows itself.

The important thing to notice here is that the value created by the worker's labor must be greater than the wages he receives for his labor-power. Otherwise, there's no reason for the capitalist to buy that labor-power at all.

Applying this example to the general formula for capital, we see that the capitalist uses $63.54 (M) to buy the commodity called "one day's worth of labor-power" to make the shoes (C), and then receives $101.66 when he sells them (M^1).

If the worker works the next day for $63.54 as well, and the shoes sell for $101.66 again, then the initial $63.54 (M) has produced two days' worth of surplus-value $76.24, resulting in a total sum at the end of $139.78 ($M^1$). At this point, he could afford to hire two workers. And if he does, the surplus-value produced in a day would double. It's through this kind of incessant movement that capital steadily increases itself.

> By virtue of being value, it has acquired the occult ability to add value to itself. It grinds forth living offspring, or at least lays golden eggs. (255)

No one has the power to stop the process by which human labor, transformed into value, increases value exponentially.

And this "occult ability" is what causes workers to be exploited.

THE DIFFERENCE BETWEEN LABOR AND LABOR-POWER

Even though the worker created it, the $38.12 of surplus-value belongs to the capitalist. This absurdity, which is known as exploitation, is justified through the logic of capitalism. Marx explains why exploitation occurs by distinguishing between labor and labor-power.

Labor-power, the ability to perform labor, is a commodity. The worker sells his labor-power to the capitalist, who then has a right to his work.

But at the point when a worker decides to sell his labor-power for $63.54 a day, or alternatively, when a capitalist decides to buy that labor-power at that price, *labor* has not yet taken place. Labor only occurs at the moment when the capitalist starts to *use* the labor-power he bought as a commodity—that is, the moment he makes the worker perform labor. And this labor is what ultimately produces the new value (M^1) of $101.66.

During this process, the labor-power priced at $63.54 a day already belongs to the capitalist, and he's free to do with it what he likes. He could even decide not to use it at all and let it go to waste, or he could just as easily use all eight hours and produce $101.66 of value. The point is, once labor-power is sold, the capitalist owns both the power to decide what to do with it *and* any surplus-value created through the labor—the worker owns neither.

The capitalist is not buying labor from the worker, but rather labor-power. Wages are paid as the price for labor-power, as it is a commodity. The worker is not being deceived in this process so the capitalist can pay him less than he's worth. Rather, if we're only looking at the exchange as it occurs in the marketplace, the guiding principle underlying the free exchange of commodities is being respected: Labor-power is being exchanged for money, ultimately making the exploitation of surplus-value difficult to see.

That said, the aim of Marx's *Capital* is *not* to expose exploitation's existence. This is a common misreading of the text. What we should be asking instead is why, if the existence of exploitation is already known, do workers continue working so obediently?

For now, the simple answer is that they have no way to make the money they need to survive other than by selling the only commodity that workers have at their disposal: their labor power. The worker owns neither the means of production nor the means of survival.

The price for their labor-power—the wage—is only barely enough to survive on. Even now, we continue to work tirelessly every day only to stop by the supermarket on the way home from work and search through the prepared meals, looking for one that might have a discount coupon stuck to it so we can afford to enjoy a canned highball with dinner.

THE TRICK BEHIND THE PROLIFERATION OF LONG WORK HOURS

The worker in the scenario we have just outlined is creating $12.71 of value with one hour of labor. If the capitalist hires that worker at a wage of $63.54 per day but only makes him work for five hours, the value produced by that labor will only be $63.54 and the capitalist's profit will be zero.

However, any labor performed after those five hours are up becomes surplus-value, which is possessed solely by the capitalist. Therefore, an eight-hour workday produces $101.66 of value, of which $38.12 is surplus-value.

But what if the capitalist wants to make that value increase even more?

The easiest solution would be to make work hours longer without raising the daily wage, right? If the capitalist makes the worker work for ten hours a day, the surplus-value would rise by two hours' worth—$25.41. If he makes the worker work for twelve hours, the surplus-value rises by $50.83.

Since the daily wage remains consistent at $63.54, the capitalist would reap $50.83 of profit for the same cost. Marx calls this kind of extra surplus-value attained simply by extending the workday, involving no extra work on the part of the capitalist, "absolute surplus-value." This is why Japan and the United States have experienced a proliferation of overlong work hours and unpaid overtime.

> The prolongation of the working day beyond the limits of the natural day, into the night . . . only slightly quenches the vampire thirst for the living blood of labor. Capitalist production therefore drives, by its inherent nature, toward the appropriation of labor throughout the whole of the 24 hours in the day. (367)

It's not only unpaid overtime—positions that are "management in name only," for example, also make people work longer hours without additional compensation; systems like "fixed" overtime salaries end up producing the same result.

As the workday (that is, the number of hours worked during a day) gets longer, the worker's lifestyle naturally becomes more difficult. Free time and time spent with family and friends disappears; there's no way to recover from the exhaustion of work, no time to pursue other interests or pastimes. Workers simply work at the expense of their health and well-being. In extreme cases, they may even work themselves to death.

LABOR-POWER IS ALSO WEALTH

Labor-power is a basic form of societal wealth. We could use this wealth to enrich everyone's lives, to help people realize their dreams, and to support society as a whole; it could be used to instill happiness and satisfaction in the hearts of working people.

But capitalism encloses the wealth represented by labor-power by turning it into a commodity. The worker's quality of life, his dreams and sense of purpose, are not relevant in the capitalist's quest to make the best use of the labor-power they've transformed into a commodity and then purchased from the worker. The capitalist's overriding concern is how much value labor-power can produce. He therefore seeks to control labor so as to maximize this production of value.

We are supposed to work in order to live, but capitalism has inverted this equation to the point that we end up living in order to work. The enclosure of labor-power's wealth through commodification leads to the underdevelopment of workers' skills as they're used up by capital.

It's not only adults who fall prey to the "vampire thirst for the living blood of labor." In volume 1, chapter 10 of *Capital*, called "The Working Day," Marx describes in detail tragic cases of children who, in pursuit of wages, stopped going to school and went to work alongside adults, often becoming addicted to cigarettes and alcohol and dying early deaths.

I said previously that *Capital* can be difficult to read, but "The Working Day" chapter is more journalistic than philosophical. Marx uses this chapter to describe the terrible fate of workers with immense detail because he truly recognizes and empathizes with their plight. Scholars explicating *Capital* often don't focus on this chapter since it is not terribly complicated at the level of theory and argument, but it's indispensable for understanding Marx's overall motivation and purpose for writing it at all.

THE RECURRENT TRAGEDY OF DEATH BY OVERWORK

Capitalist production always seeks to wring the most labor-power possible out of workers, leading to their immiseration, both body and soul, as well as the degeneration of their skills and quality of life.

Marx writes in the chapter "Labor Day" in *Capital* about a June 1863 incident that made all the dailies in London at the time.

> [The articles] dealt with the death of the milliner, Mary Anne Walkley, 20 years old, employed in a highly respected dressmaking establishment, exploited by a lady with the pleasant name of Elise . . . These girls work, on an average, 16½ hours without break, during the season often 30 hours, and flow

> of their failing "labor-power" is maintained by occasional supplies of sherry, port, or coffee . . . Mary Anne Walkley had worked uninterruptedly for 26½ hours, with sixty other girls, thirty in each room. The rooms provided only 1/3 of the necessary quantity of air, measured in cubic feet. At night the girls slept in pairs in the stifling holes into which the bedroom was divided by wooden partitions. And this was one of the better millinery establishments in London. (364–5)

The title of the newspaper article was "Death by Simple Over-Work." In other words, the same phenomenon known as *karōshi* in Japanese. The same tragedy that befell Mary Anne and described 150 years ago in *Capital* is still happening today across the world, in the Global North and South alike. There's no way, unfortunately, to write this part of *Capital* off as a relic of the past, no way to reassure ourselves by saying, "Oh, things were really terrible back then, weren't they?"

Consider, for example, a 2008 incident of *karōshi*—death by overwork—that occurred at the restaurant chain (izakaya) Watami. A woman killed herself a mere two months after being hired there; during those two months, she'd been forced to work over 227 hours outside regular work hours. Despite being promised workdays limited to eight hours and two days off per week, once on-site she was told, "Every hour we're open is a work hour." Not only was she made to work many hours beyond what she'd been promised, but her days off were filled with "voluntary" work activities, report writing, and tests of how well she'd memorized the company's "management philosophy."

In 2015, the major advertising agency Dentsū saw a similar incident, in which a woman who'd graduated from Tokyo University committed suicide from overwork during her first year at the company. Long work hours at the office had become routine, and the woman in question had regularly slept only two hours a day, which totals only ten hours of sleep a week.

Our awareness of workers' rights and attention to improving workplace conditions should, by all rights, be better now than in Marx's time. Yet the pressure on workers to work longer and longer hours hasn't lessened at all, and so labor-power as societal wealth continues to be destroyed.

WORKERS DRIVEN BY "FREEDOM"

Why is it, then, that workers don't resist exploitation, even to the point of being driven to suicide from inhumanly long work hours? Couldn't they just walk off the job, or resign?

According to Marx, workers cannot escape because they are "free." But what does that mean? Why would being "free" prevent escape?

Marx argues that workers in a capitalist society are "free" in two senses. One is that they are not forced into labor in the same way that, for example, enslaved people are. This is not a society with enforced caste systems based on heredity or family vocation, so in principle, anyone can do whatever job they like wherever they please.

Yet, while we may be free of enslavement and caste, we're also "free" from accessing the means of production. This means, in practice, that we're blocked from participating in the production of the things we need to live. In other words, we're free of access not in the sense of *unfettered*, but rather in the sense of *without*. This is a consequence of the enclosure of the commons discussed in the previous chapter.

Once cut off from the means of production, most people cannot live self-sufficiently; most of us have no choice but to buy commodities, which means we need money to survive. And in order to get money, we need to have something to sell. For the average person, the only thing they have to sell is their labor-power.

Workers in a capitalist society, unlike enslaved people, can sell their labor-power "freely." In other words, the relationship between workers and capitalists is, in principle, free and equal, at least until a contract is signed—workers can theoretically sign contracts with any company they wish. But this is the full extent of their freedom. After that, work and slavery become more difficult to differentiate.

How exactly does this work? The Marxist economist Yoshihiko Uchida explains it this way in his book *The World of Capital*:

> Workers can dispose of their labor-*power*, but they have absolutely no power to dispose of their *labor*. If you think I'm wrong, just try freely disposing of your own labor at your workplace—all that will end up disposed of is you yourself . . . The total control over the ability to dispose of one's labor-power and the total lack of control over the ability to dispose of one's labor are two sides of the same coin.[4]

The "ability to dispose of one's labor-power" is the ability to decide whether or not to sell it. This is the power workers possess. But as soon as that labor-power is sold, the ability to dispose of your *labor*—that is, the freedom to choose *how* you work, or don't—is completely lost. The moment the contract is signed, the worker is subject to the supervision and commands of the capitalist. If you ignore these commands and work as you like, you'll simply end up out of a job.

The capitalist decides both how workers work and whether they receive any value produced by that work. At the site of labor, neither freedom nor equality exist. As the great labor scholar Makoto Kumazawa memorably put it, democracy stops at the factory gate.[5]

Even knowing this, though, in a society where anything and everything has been commodified, workers are compelled to "voluntarily" give up their freedom in order to live. There's no real choice here. This is why Marx compared the worker's plight to that of an enslaved person, calling them "wage slaves."

But we don't want to accept the fact of our own enslavement. We want to think of ourselves as free. (This is why we trumpet so insistently our ability to buy whatever we want in the marketplace.) Capitalism takes advantage of this feeling to drive us to work harder, pushing ourselves to our very limits.

WHY DO WE PUSH OURSELVES SO HARD?

It's true that workers have the "freedom" to quit their jobs and escape an exploitative and immiserating work environment. So why don't the Mary Anne Walkleys of today just walk off the job? They have their lifestyles to maintain, of course, but there is also fierce competition with other workers, and so to the average worker it appears more feasible to work at even the most exploitative workplace. But Marx points out that there's more to it than even that—the special magic of capitalism is at work here as well.

Enslaved people in precapitalist times were bought and sold by forces completely external to them, their rights and personhood disregarded as they were moved from place to place like livestock. They could not escape, as they would face horrific consequences if they tried. They labored against their will, driven by terror.

Yet enslaved people generally had some of their most basic needs met. In the same way livestock is taken care of, enslaved people were maintained by their owners because of their worth as property.

Contrastingly, in a capitalist society, no one's most basic needs are guaranteed. Capitalism frees societal wealth from the commons in the same way it frees people from feudal serfdom and other master-slave relations and communal social arrangements. But to be "freed" from communal society means also being free of, or rather cut off from, relations of mutual aid and care.

That's why today, even if you are employed, if you end up losing your health or your job, you can quickly find yourself unable to make ends meet. Marx refers to workers, as people always facing this sort of risk, as the "potentially" poor. You might recall the Occupy Wall Street movement—tent cities famously set up in New York City's Zuccotti Park in the heart of the financial district in the aftermath of the 2008 financial crisis. The movement criticized inequality and protested the United States' weakened social safety net; once you lose your job, it's all too easy to suddenly require public assistance.

Workers in a capitalist society only have the commodity called labor-power to help them navigate this unstable environment. They must decide for themselves who to sell it to in order to live. This pitfall is known as "personal responsibility."

> The slave only works under the impulse of external fear, but not for *his own existence*, which does not belong to him, and yet it is *guaranteed*. The free worker, in contrast, is driven by his wants. The consciousness (or rather the *notion*) of free self-determination, of

> freedom, makes the one a much better worker than the other, as well as the feeling (consciousness) of responsibility bound up with this.[6]

Workers are not only motivated by the fear of losing their livelihoods. They're even more motivated by the pride they take in having, by their own volition, made the decision to work. Workers feel a sense of duty when doing their jobs. Many of us surely have memories of voluntarily handing over our freedom during a job interview, promising that we'll give a company absolutely everything we have. Here, we see an obvious difference between workers and enslaved people, who may have their minimum needs met but are compelled to work against their will.

Workers who apply themselves to their jobs out of a sense of personal responsibility work harder and more intentionally than enslaved people working against their will. They're usually motivated to do a good job and often blame themselves when they make mistakes. They take responsibility even when their superiors make absurd or impossible demands. This is a capitalist's dream come true, evidence that workers have internalized the logic of capital, a mentality that thinks always about how best to benefit the *capitalist*. The political scientist Satoshi Shirai has called this "the subsumption of the spirit."

At base, the concerns of the capitalist, who seeks the unlimited increase of value, and those of the worker, who seeks a human, livable existence, can never truly align. Yet the "free," voluntary worker imagines that the type of worker the capitalist desires is an ideal model to pursue—the portrait of an ideal person. An illustrative example of this prototype is the popular figure of the *mōretsu shain*—the ultimate go-getter, willing to sacrifice everything for the good of his company—popularized during the postwar era of high-speed economic growth, and during the bubble economy of the 1980s, through the energy drink slogan *Can you keep up the fight for twenty-four hours?*

Marx warns us that in a capitalist society, a worker's sense of personal responsibility and drive for self-improvement may constitute a type of subjectivity, but it's in constant danger of being subsumed by the logic of capital.

DON'T RAISE WAGES, SHORTEN THE WORKDAY!

Workers subsumed by the logic of capital often sacrifice their health and family for the sake of work, begrudging even the time they spend sleeping. Technological advances have raised production power astronomically, yet the modern worker still works more than the average worker did in Marx's time.

But even with everyone working so much, Japan has been unable to recover from the long-term economic recession that began when the bubble economy burst at the end of the 1980s. Funding for research has been cut, and innovation has stalled.

Underpinning this situation is market saturation. The time has passed when every household needed to stock itself with appliances, televisions, cars, and so on. Demand has been reduced because people now just buy replacements for what they already have. International competition has made it difficult to find new global markets to replace domestic demand.

The quickest, easiest way for industry to raise profits in this situation is to either cut wages or lengthen the workday. Overall sales over the past decade have remained stable even as profits have continued to rise. Industry writ large has united to cut costs through drastic anti-worker measures like lowering wages.

If a daily wage that used to be $63.54 drops to $50.83 while the eight-hour workday remains the same, the "surplus worktime" that produces surplus-value increases by four hours. Simply put, cut wages go directly into capitalists' profits.

Let's look at a true wage cut—no more overtime pay, no more bonuses, forced early retirement leading to reemployment at lower pay. The next steps are even more serious, and include decreases in the number of permanent staff who are then replaced by nonpermanent staff or even temps.

It's of course important to try to raise wages that have been unfairly cut. Wages have always been a powerful rallying cry for labor movements and in negotiations between labor and management.

But interestingly enough, Marx asserts that it's more important to "limit" (that is, shorten) the working day than it is to raise workers' wages, even as wages in his time were lower than they are now.

> The limitation of the working-day is a preliminary condition without which all further attempts at improvement and emancipation must prove abortive. (415)

In effect, Marx is saying that raising wages is meaningless if the problem of long work hours remains unresolved.

If the capitalist accepts the call to raise wages, this of course lessens the extent of exploitation. But this will also lead those workers living in a capitalist society to want to work even harder, a development that would surely be welcomed by the capitalist. If higher wages can make workers "voluntarily" work longer hours, the surplus-value—the profits—reaped by the capitalist will, counterintuitively enough, increase.

The aim of the capitalist is to enclose the societal wealth represented by labor-power by commodifying it. This enclosure consists of the theft of free time. If workers put in longer hours for higher wages, this only preserves the system of wage slavery.

Even if a worker's wage rises, if his time is still being taken from him, he'll remain unable to enjoy any leisure time playing with his children or pursuing his hobbies.

If you're too busy to cook, you can buy commodities like takeout food. If you have no time to hang your clothes to dry, you can always buy a dryer. There are autonomous vacuum cleaners, too. Many people simply hire a cleaning service.

But none of these convenient services actually offer us any more free time. We must work even longer, in fact, just to be able to afford them. In this way, the unlimited expansion of the workday paves the way for expanded commodification, along with new business opportunities for the capitalist.

WE MUST TAKE WEALTH BACK FROM CAPITALISTS

Marx emphasized the importance of shortening the workday because he saw it as a direct way to return wealth to the people. To protect the richness of our daily lives, we must refrain from turning our labor-power into a commodity, or at least put limits on the extent to which we do so. The easiest and fastest, as well as most effective, way of doing this is not through raising wages but through "the limitation of the working day," as Marx put it. Of course, shortening the workday becomes meaningless if wages go down as a result, so if we consider wages as an hourly rate, they will necessarily have to go up.

If workers wish to protect their way of life while acquiring the knowledge and skills it takes to realize their dreams, they need to limit how much their labor-power and lifetimes end up harvested as commodities. However, as long as they are compelled to deal with capitalists who desire nothing more than to enclose labor-power as a commodity, there is no real way to prevail on an individual level. Workers must band together—as we'll see in chapter 5, Marx calls this "association"—to defy capitalists. This takes time: time to gather together, to talk things out, to study and do research. For this reason, too, the workday must be shortened.

TWO OPPOSING MOVEMENTS

Right now, the world is moving in two directions on this issue. The prevailing trend has been to extend the workday beyond its traditional hours.

Unlike in Marx's time, the site of labor today has expanded far beyond the workplace. In many cases, as long as you have a laptop and network connection, you can work anywhere. Telecommuting became common during the COVID-19 pandemic, and now many meetings are conducted remotely and people take so-called "workations" in which they travel to luxury resorts while continuing to do their jobs.

The blurred distinction between work time and private time has resulted in an overall lengthening of the workday. If the phone rings, it doesn't matter if you're having dinner with your family, you're expected to snap into work mode. If you wake up in the middle of the night, there's always an email from a different time zone to respond to on the phone you keep by your bed. Working from home means you have no commute, but it also means that the hours when the trains run no longer function as limits on the workday, effectively making you available to work all twenty-four hours of the day.

And that's not all. When we use platforms like Google and Facebook, the data we share becomes transformed into commodities that produce value for those companies. They collect our data and sell it to companies that use it to advertise their products to us and profit.

We imagine ourselves using a site like Facebook in our private time, for our private pleasures, but as we upload photos onto Facebook or search for things using Google, we're busily working to produce the commodity called "data" that these companies—sometimes collectively referred to as GAFA (Google, Amazon, Facebook, Apple)—need to make money. And we do it all for free!

The addictive qualities of games and social media make us, as modern men and women, unable to put down our phones, and so we work and consume every hour of the day we aren't asleep. If self-driving cars become the norm, soon we'll have that time, too, to devote to our phones, and the giants of the tech world will only amass more wealth. We're told we should be glad that the world has become more convenient, while in truth, our entire lives have become subsumed by capital—this is the end result of the smartphone addictions that have given rise to what might be called the "digital proletariat."

TOWARD A SHORTER WORKDAY

There's another movement, though, emerging these days—the movement to shorten the workday, not lengthen it. The former Finnish Prime Minister Sanna Marin famously proposed a limitation on work hours. Elected in 2019 as the world's youngest prime minister, Marin had long advocated for a four-day workweek and a six-hour workday; once in office, she made these changes an explicit goal of her administration.

Iceland also experimented with a four-day workweek, and the data showed no reduction in labor productivity. In the UK, too, they've begun experimenting with a similarly shortened workweek. In Belgium, workers can choose to take three days off a week, though on the condition that they work longer workdays. This movement continues to grow even as the expansion of remote work proliferates as well.

This shouldn't come as too much of a shock. The eight-hour workday was already on its way to becoming a norm in Marx's time. Isn't it more of a mystery why, after the passage of 150 years and the dramatic increase in productive power over that time, we work the same number of hours—or even more!—that people did back then?

We must ask ourselves, what's wrong with a twenty-hour workweek, anyway? In terms of its impact on societal wealth, nothing. But in context of capitalism, it would be disruptive. It would stop the valorization of capital in its tracks.

Unfortunately, we don't see any movement in Japan or the United States challenging the tenets of capitalism by limiting work hours. On the contrary, we're seeing a rise in the villainization of those who need public assistance, an increasingly strident moralism that insists, "If you don't work, you don't eat!" People are forced to take on second jobs and then fill their days off work with self-improvement seminars to get ahead. We're selling more and more of our time to other people. But what are we getting out of it?

Exploitative so-called "black" corporations recently became a widely discussed social problem, yet Japan has still done nothing to solve its overwork problem. The structural problems underpinning the phenomenon must be addressed for us to be able to demand a new way to work. Marx's insight into the problem of the working day, as introduced in this chapter, provides us with an important clue as to how we might begin this process.

3

INNOVATION BREEDS BULLSHIT JOBS

KEYNES'S OPTIMISM, REALITY'S PESSIMISM

Why didn't long work hours disappear under capitalism? As we saw in the previous chapter, rather than shortening the workday, the movement of capital literally drives people to work themselves to death.

About sixty years after the publication of *Capital*, the British economist J. M. Keynes predicted that the development of capitalism would result in the shortening of the workday. He thought that by the year 2030, work would shrink to about fifteen hours a week, and people would enjoy all the free time they could ever want. The biggest problem of the twenty-first century would not be how long people work or the nature of their working conditions, but rather finding ways to spend all the free time they'd have on their hands.[7]

A hundred years have passed since then, and our lives have indeed changed in many ways. All sorts of technological innovations have occurred that neither Marx nor Keynes could have imagined in their wildest dreams. The ability of society to create things—that is, its productive force—has risen dramatically, leading to similarly dramatic high-speed economic growth.

In his essay, Keynes states that "from the earliest times of which we have record—back, say to two thousand years before Christ—down to the beginning of the eighteenth century, there was no very great change in the standard of life of the average man living in the civilized centers of the earth."[8] And indeed, worldwide GDP had remained more or less even until about 1800.

With the rise of capitalism and the accompanying development of various technologies, though, the worldwide GDP increases dramatically. This is especially true in the period after World War II; the graph's line jumps almost straight up. In the last thirty years, internet technology and smartphones have become ubiquitous, and there have been notable advances in robotics and AI research.

Given all this, it would seem that Keynes was correct—those of us in the Global North should be living lives of leisure, working almost not at all! And yet we know this is not the case. Indeed, we're driven to work harder and longer than ever, living in fear of the "rise of the robots."

This general unease was best expressed in the fall of 2013, when Oxford University professor Michael A. Osborne presented a report called "The Future of Employment." Analyzing over seven hundred jobs in fine detail, Osborne estimated that almost half of US workers would, in ten to twenty years' time (that is, starting in 2023), lose their jobs due to advances in technology. And these weren't just factory jobs, either—high-salary, high-status jobs like accountants and bankers were also on the list, contributing to the report's impact.

Osborne's findings agree with Keynes in that both predict a major reduction in human labor. But Osborne didn't envision a future utopia in which people would be free not to work if they didn't want to—rather, it was a future dystopia in which people would end up unemployed and then, unable to find work, would be unable to survive. So, we end up spending our lives working harder than ever, driven by the fear that productivity levels will rise exponentially to the extent that humans will become entirely obsolete.

In truth, technological advances have made many jobs very tedious. We work ceaselessly at low paying jobs that nearly anyone could do out of fear of being fired. Unable to put away money for the future, we end up stockpiling fears instead.

Capitalism may seem as though it produces huge amounts of wealth, but we find ourselves barely scraping by anyway. Our desire and passion for life withers as a result, and we become spiritually and emotionally poorer for it. Over 180 years ago, Marx, then just in his mid-twenties, came up with a term for this phenomenon: alienated labor. Filled with righteous anger, Marx described the alienated worker:

> Light, air, etc.—the simplest animal cleanliness—ceases to be a need for man. Filth, this stagnation

> and putrefaction of man—the *sewage* of civilization (speaking quite literally)—comes to be the *element of life*—for him. Utter, unnatural depravation, putrefied nature, comes to be *his life-element.*[9]

Shouldn't work be a more attractive thing, something that makes your life richer, not poorer? If anything, Marx's plea rings all the truer today.

We exhaust ourselves at tedious jobs, then go home to our tiny apartments and wash down unappetizing late-night convenience store meals with a beer as we scroll through YouTube or Twitter—doesn't something seem off about this? Having a "case of the Mondays," counting the days till the weekend or a short vacation—these are feelings of alienation we all recognize, surely.

I would like to take a closer look now at Marx's analysis of alienation in *Capital*. (Those familiar with the work of Wataru Hiromatsu or Louis Althusser might be surprised to hear that *Capital* contains such an analysis, but I assure you it does.) After that, we will consider some ways to overcome it.

THE PRESSURE TO CHEAPEN EVERYTHING UNDER CAPITALISM

So why is it that contrary to Keynes's prediction, the rise in productive force didn't free workers from work or make them happier? According to Marx, the key to understanding this is to consider why productive force rose so much under capitalism in the first place—in other words, why capitalists seek to raise the forces of production as much as possible at all times.

The reason capitalists always want to raise productivity is, to put it bluntly, so that they can make commodities as cheaply as possible and beat their competitors in the marketplace.

If you can produce something cheaply without lowering its price, profits go up. Tangentially, if you do decide to sell the item for less, this enables you to expand your share of the market. As your share of the market expands, the principle of low profit margins with high turnover takes over, and you'll continue to make money.

As we saw in the previous chapter, capital is the movement to increase value, while capitalism is the pursuit of endless profit-making. If increasing the forces of production can make production cheaper, the resulting increase in profits incentivizes capitalists to do all they can to raise their levels of productivity.

For example, let's say some capitalists are producing hand-sewn shirts that sell for $63.54 each. One day, one of the capitalists—let's call him Yamada—imports a sewing machine from overseas that helps him successfully cut enough costs that he can make a profit even if he sells a shirt for $6.35. He could go ahead and sell that shirt for $6.35, of course, but the other capitalists are selling the same shirt for $63.54. So Yamada decides to raise the price of his shirt to $31.77—$25.41 more than the initial $6.35 price. Now, as the retailer with the most competitive price for shirts, he's able to dominate the marketplace and make the most profit.

Those extra $25.41 naturally all end up in Yamada's pocket. Marx calls the difference between the market price ($31.77) and the price that properly reflects the amount of labor necessary for production ($6.35) "extra surplus-value" (436).

As productive force rises, profits grow exponentially even if products are sold more cheaply than before. Furthermore, a shirt that costs half the market price will fly off the shelves. Of course, the other shirtmakers cannot abide this situation without responding. Soon everyone will import the new machines and start selling shirts at $31.77, too—or, better, they'll try to steal back Yamada's share of the market by selling them at $25.41. In the end, everyone will be using the same level of technology to make their shirts, and the price will be driven down to $6.35: The extra surplus-value disappears.

What happens to capitalists who fall behind in the race to lower prices? They end up with leftover stock made in the old $63.54–costing way. Rich people may fixate on the difference between a handmade shirt and machine-made one and pay the higher price—but such people are the vast minority of consumers, and overall demand will still fall. On the other hand, if these unlucky shirtmakers decide to sell the shirts at the new, cheaper market price, each sale will put them further into debt and they will soon go bankrupt.

> Capital therefore has an immanent drive, and a constant tendency, toward increasing the productivity of

> labor, in order to cheapen commodities and, by cheapening commodities, to cheapen the worker himself. (437)

As long as the capitalist remains a capitalist, he must always strive to produce his goods more cheaply and efficiently than his competitors in order to make money. Consequentially, the capitalist strives to cheapen commodities, or at least raise productivity levels, at all times, which in turn drives technological advancement and innovation. The biggest factor, therefore, contributing to the dramatic rise in productive force that occurs under capitalism is extra surplus-value.

THE RELATIVE SURPLUS-VALUE CREATED BY STRENGTHENING FORCES OF PRODUCTION

Once price wars spread into society as a whole, commodities of all sorts become available cheaper than ever before. You might think, *So I can buy whatever I want for less money? Wonderful!* But this development may not be as welcome as it might seem at first. As Marx put it in the previous quote, the endpoint of this process is to "cheapen the worker himself." How does that happen?

A worker becomes "cheapened" when labor-power cheapens, which takes the form of lower wages. To understand this, we must first understand how wages are determined.

Wages are not determined solely through the relation of supply and demand in the marketplace. For example, although the Japanese government is increasingly concerned about the perceived lack of manpower to fill jobs, wages have remained stagnant. Marx states that in fact, wages are dictated by how much money workers need to sustain themselves.

A worker who gets tired over the course of a workday must eat dinner and get a good night's sleep in order to work again the next day—this is how workers renew the commodity known as labor-power so they can sell it again. Marx calls this the "reproduction of labor-power." The cost of the reproduction of labor-power is determined by how much a worker will be required to work so he can produce the necessities to get by in his daily life.

What these necessities actually are varies by time and place. Food, shelter, clothing, leisure—these things are obviously necessary, but in today's global society, so are smartphones and internet connections. Many places necessitate owning a car. Workers' long-term needs include educational fees for children and savings for retirement. All of these things and more go into the calculation of wages.

Let's say a daily wage of $136 (this represents eight hours of work) is the amount a person needs to get by. As productive force increases, things like fast fashion and fast food lower the cost of living, and the same person can now have the same quality of life for a daily wage of $102 (which for our case is representative of six hours of work). Seeing this, the capitalist realizes that he can lower the daily wage to $102. In practice, this translates into the rise of precarious employment practices like hiring more workers on only a temporary basis so as to cut costs overall.

Raising the productive force and making the cost of living cheaper doesn't change the value of the labor a worker performs in an hour. If that labor produces $34 of value and the worker still works the same eight-hour day as before, then lowering the daily wage from $136 to $102 translates into $34 more surplus-value for the capitalist.

Marx calls this type of surplus-value, created through the cheapening of labor-power, "relative surplus-value."

Why "relative"? As we saw in the previous chapter, the production of absolute surplus-value occurs when the length of the workday is extended, which makes the absolute value of that day's production increase. By contrast, the production of relative surplus-value entails no change in the length of the workday and no change in the value of that day's production. Rather, the trick to raising profits lies in lowering the value of the labor-power itself.

This concept may seem a bit complicated, but the main point is that productive force has been rising precisely due to capitalists' desire for more relative surplus-value. It has nothing to do with wanting to "liberate humanity from labor" or "enrich the lives of workers." Lowering prices isn't done to benefit consumers. Rather, it helps drive workers to perform increasingly exigent, exploitative forms of labor.

INNOVATION FOR WHO?

We've seen how productive force increases under capitalism so that capitalists can maximize the amount of surplus-value they attain. The race to lower prices and the logic of low profit margins on high-turnover sales are also relatively intuitive to understand.

But the technological advances that have raised productivity—in other words, the innovations sought by capitalists—don't just increase value, they also enable further domination of workers. In fact, this is Marx's most pressing concern regarding the increase of productive force under capitalism.

Capitalists seek extra surplus-value by making workers work "more efficiently" so that the commodities they produce can be made as cheaply as possible. In other words, they seek to increase their productivity. The important thing to remember here is that efficiency has nothing to do with making the worker's life easier.

Under capitalism, new technologies aren't developed and introduced with the aim of freeing workers from arduous labor or complicated tasks. Rather, new technologies are introduced to more efficiently control and surveil workers, and to guarantee that they will work tirelessly and without complaint. It's here that Keynes's prediction falls the shortest in that it failed to foresee how technological innovation causes people to work more, not less.

Marx points out that as the forces of production increase, workers, lives do not become easier; rather, workers become subsumed by capital and lose their agency, eventually becoming capital's slaves.

But how is the rise of productivity levels connected to capital's increased domination of workers? To answer this, I'd first like us to recall the discussion of metabolism from the beginning of chapter 1, specifically how humans manage their metabolic relation to nature through labor conceived "as a conscious activity with a clear purpose."

This kind of conscious, purposeful labor is made up of two elements: conception and execution. (These are terms used not by Marx himself, but rather by the excellent Marxist scholar Harry Braverman in his book *Labor and Monopoly Capital.*)[10]

For example, let's say you think to yourself, "I want something hot to eat. Therefore, I need a tool that will help me cook food." You will then use your intellectual capacity to think about the form such a tool should take: what it should be made out of, how durable and heat resistant it should be, and so on. This is *conception.* Let's say that in this case, the process of conception results in the decision to make an earthenware cooking vessel.

Now you must use your hands to actually create the vessel according to your conception. To make a sturdy, functional piece of pottery, you must find and dig up suitable clay, add the proper amount of water to it, shape it correctly, and fire it.

The chain of tasks needed to make the conception of a product into a reality is *execution*.

Marx himself calls the work of planning things in one's head "mental labor," and the work of realizing these plans with one's own hands "manual labor." It's important to remember that the distinction between mental and manual labor should not be confused, as it often is, with the distinction between "white-collar" and "blue-collar" work. Originally, all human labor was inextricably made up of both mental and manual labor.

Marx argued that under capitalism, increasing productivity entailed the separation of conception and execution—or, to use Marx's own terms, mental and manual labor—in the labor process. Conception became monopolized by capitalists or by site foremen hired by capitalists, while workers became responsible only for execution.

Artisanal labor like pottery making is an easy-to-understand example of a type of labor that unites conception and execution. Artisans are people who have acquired skills and knowledge through years of training and are thus able to use their judgment to take charge of the whole job, from deciding what to make (conception) to using their own hands to make it (execution).

Well-seasoned artisans can tell how hot the kiln is by the color and flickering of the flame; they can determine the proper firing temperature and how long to fire the clay by taking into account the day's temperature and humidity, the state of the clay, and so on. It's impossible to turn this kind of know-how into a set of written instructions or to buy it with money. Work processes and materials vary widely between production sites. The only way to transmit this kind of knowledge is through apprenticeship. At the time capitalism began its rise, most items were made by artisans and experienced craftspeople, and in the larger cities of Europe, they formed guilds—trade associations known in German as *Gildes* (for merchants) and *Zunfts* (for craftspeople).

A guild was made up of craftspeople and artisans who were devoted to the same craft and took on apprentices to mentor. Together they secured raw materials, shared skills and techniques, and set prices, resulting in strong bonds between these tradespeople. Guilds had many rules and regulations. For example, you couldn't take on apprentices until you'd been an apprentice yourself for a certain number of years, or if you set up your own shop, there were limits to how many apprentices you could take on. The amount of profit you could make was also set, and advertising and discounts were prohibited. Those who broke these rules lost their ability to be a master with apprentices, and sometimes suffered even more extreme punishments, including having their arms broken.

Looked at from today's perspective, this surely seems like an oppressive, restrictive system. But these rules didn't arise because artisans were stubborn or narrow-minded. For them, the point of production was not to maximize profits, but rather to create a stable order in which everyone's livelihood could be protected. Managing their ability to both conceive and execute their craft themselves, these artisans avoided excessive competition and protected their ability to control the conditions and environment in which they worked.

Capitalism, however, cannot tolerate such a system. The capitalist will order the artisan to make one hundred pots in three days, and the artisan will refuse, saying, "That's impossible—I need a week to do that." The capitalist might say, "You don't need to put a pattern on it, it would be cheaper if you didn't," and the artisan will respond, "This is my craft and my method, I won't change it." If you anger an artisan, he will no longer make anything for you at all, so the capitalist is at the mercy of his wishes. This is the state of things when capital is forced to depend on the skills and knowledge of people in command of both the conception and execution of their labor.

As we've seen, capitalists always strive to produce as much surplus-value as they can in as short a time as possible. They also want to increase the number of workers in their employ as well. Pressed by market competition, they cannot wait for new artisans to emerge from yearslong apprenticeship systems.

Therefore capitalism thoroughly dismantled the guild system. But how to prevent artisans from resisting? The answer lies in the separation of conception from execution in their labor. In other words, by reorganizing the technological conditions of labor, capital was able to transform the means of production to its own advantage, refitting the labor process to maximize the creation of surplus-value.

DISEMPOWERING WORKERS THROUGH THE DIVISION OF LABOR

So how can conception and execution be separated?

The easiest way to do it is to divide the labor process into its constituent parts and give them each to different workers. This means that it's the capitalist who understands the whole process by which, say, a pot is made: what tasks are involved in each step, what tools are needed and how they're used, how long each step should take, and so on. In this way, the specific methods used by an artisan become standardized and transformed into a series of simple tasks.

Even if you can't make a pot by yourself, certain parts of the process can be turned into a uniform set of instructions and performed by even a novice with only a minimal amount of training: working the clay, for example, or placing it in or taking it out of a mold. Each step becomes its own menial task, and it also becomes easy to create special tools to make each task even simpler to perform.

Turning the workforce from a band of accomplished artisans into a group of novices with a minimal skill set makes it much easier for the capitalist to direct and control the production process. Once the different tasks have been standardized, production can dramatically speed up. As the production process becomes more uniform, the division of labor makes it easy to improve efficiency. When menial tasks become the focus of employment, one can hire more workers, which in turn enables the scale of production to expand as well.

There are, of course, still some things only artisans can produce even now. But these are commodities that never lent themselves to capitalistic production anyway. Capitalists don't want to make high-quality handmade pottery—they want to make things cheaply that are *good enough*, things produced in high enough quantities that if you buy one and lose or break it, you can easily go buy another one at a low price. As opposed to carefully made items crafted by a single skilled artisan, these are things made by many people each doing one simple task within a workflow—things that can be produced efficiently, cheaply, and in great numbers.

The spread of capitalism put pressure on artisan guilds to lower prices within the marketplace, leading to their eventual dissolution. Unable to remain independent as producers, artisans either went out of business entirely or found a place within the division of labor overseen by the capitalists. In this way, the weakening of the guilds is directly linked to the increase of power wielded by capitalists.

As a result, the guilds lost their special status over the course of the eighteenth and nineteenth centuries, as various nations reformed their laws under the flag of free enterprise. In Marx's home country of Germany, the guilds began to dissolve in 1869, at which point unrestricted capitalism began to take shape in earnest.

THE REAL REASON WHY WORK IS SO PAINFUL

Master artisans weren't the only ones suffering from productive force's rise as the division of labor took over the production process. A cruel fate awaited the apprentices in training, as well.

These novices ended up subsumed completely within the division of labor set up by capitalists, who stole any opportunity they may have had to participate in *conception*. Novices were instead expected to repeat the same menial task given to them on the factory floor day after day, and deprived of the ability to acquire the skills and judgment required to holistically understand the work process. In this way, the movement of the market ends up steadily transforming the content of work itself. It's easy to replace a menial worker doing a simple task. This is how workers lose out on *execution* as well, since they no longer have the opportunity to acquire the rich experience that allows artisans to become experts at their craft.

The base reason workers are forced to sell their own labor-power as a commodity to capitalists is that they don't have access to the physical means of production—that is, to the ability to make what they need to live. But as Marx points out, under the system of divided labor, workers lose the ability to produce anything for themselves at all.

Workers who work for years and end up only knowing how to do one simple task are employable only under a system of divided labor. (After all, they can no longer make a complete product on their own!) This means they have no choice but to submit to the control and demands of capital to sustain their livelihoods. The division and sharing of labor thus both become reorganized along capitalistic lines, robbing workers of their subjectivity.

Furthermore, a worker doing a simple task that can be performed by anyone can always be easily *replaced* by anyone. If they don't want to lose their job, workers must swallow their resentments and dissatisfactions and strive to meet their quotas each day without complaint. This is how the master-servant relationship between the capitalist and worker gets stronger—the biggest reason why work has become so painful is that workers are no longer given the opportunity to exercise their own judgment in the workplace.

TAYLORISM ROBS US OF OUR HUMANITY

Here I'd like to introduce a prime example of a thorough separation of conception and execution in real life: the introduction of the system known as Taylorism into industry at the beginning of the twentieth century. Taylorism is a management method that was first promoted by the American engineer Frederick Winslow Taylor in his book *The Principles of Scientific Management* (1911). Taylor started as a mere clerk in a machine manufacturing plant, but soon became a foreman and then went on to become a business consultant; these days, he's seen as the father of American management studies.

Taylor's method was to separate the production process into its constituent steps, then look at the movements and actions necessary to perform each step and the minimum time it would take to complete it; from this, he calculated the time it should take to complete the production process as a whole. To minimize wasted movement as much as possible, he devised arrangements on the factory floor that optimized the performance of each necessary action according to the bodily dimensions and know-how of the worker; he also devised specialized tools to help workers perform each task more efficiently. In other words, Taylorism exemplifies how the technical conditions of production become completely transformed under capitalism.

Simultaneously, the people who plan and administer production became completely separated from those who perform the actual work. This is meant to allow the worker to concentrate solely on completing "the task at hand" within a set time limit. Those who fail to meet this time limit face punishment or dismissal, while those who complete more tasks faster are rewarded. The introduction of an inherently discriminatory system that bases wages on production volume drives workers to compete with each other to perform menial tasks as fast as possible.

What inspired Taylor to create such a system? When he first began working at the steel factory, the workers on the floor refused to follow his orders. Taylor was frustrated that the knowledge and skills he possessed were completely incompatible with the know-how possessed by the workers beneath him. In other words, he felt disrespected.

In response, Taylor devised a system by which these workers' knowledge and their skills—that is, their mastery of both conception and execution—would be broken down by forcing them to "voluntarily" compete with each other to maximize the efficiency of their work, work that was now conducted solely according to *his* wishes.

Taylor is credited as the father of "scientific management," but Taylorism is nothing more than the enclosure of the commons, which in this case are represented by workers' common knowledge and learned skills. Workers must now submit themselves to a labor system remade to accommodate capital, as it is now capital that monopolizes production-related knowledge and know-how.

Even so, it's sometimes the case that when productivity jumps temporarily, the boon of increased profits translates into higher wages for workers as well. This is why even labor unions have ended up permitting capital to steal access to conception from workers.

As the era of high-speed economic growth grinds to a halt, however, capital no longer offers workers even these sorts of leftovers. Instead, the position of workers steadily weakens, wages go down, and working hours lengthen.

Capitalism raises the forces of production while robbing workers of their autonomy and any sort of free time they might have used to enrich their lives and cultivate themselves in a human way. That's why no matter how much productive force increases, the society overflowing with free time envisioned by Keynes will never come to pass as long as capitalism rules the day.

WORKERS SERVING MACHINES

As a result of the separation of conception and execution, workers lose the ability to conceive of their own work and consequently lose their agency at the workplace. Enclosed into a series of menial tasks, human labor becomes steadily deprived of the "wealth" of know-how that characterized it in precapitalist times.

This tendency reaches its most complete form in large-scale industry, where the production process has become more mechanized. In chapter 15 of *Capital*, titled "Machinery and Large-Scale Industry," Marx writes:

> In the factory we have a lifeless mechanism which is independent of the workers, who are incorporated into it as its living appendages . . . It is not the worker who employs the conditions of his work, but rather the reverse, the conditions of work employ the worker. (548)

The "lifeless mechanism" in this instance refers to machinery—things that were produced by past labor. Marx refers to this kind of labor as "dead labor." Live humans perform "living labor," but they end up doing so as an appendage to the "dead labor" encased within the machinery. Before, humans used their tools to do work—that is, they made the tools to serve them—but at the site of production within large-scale industry, the tools use the humans, making them their servants.

In chapter 1, we looked at how we're manipulated by value in the form of commodities; in chapter 2, we looked at how we're being manipulated by the movement of capital. Here, we see how the era of industrialization has led to workers being manipulated by machines. This is what it means to say that the production process itself has become reified—that machines, as things, have switched places with workers, as humans, reversing their roles completely.

It's the machinery that hums busily along during the production process now, confronting the human worker as "dead labor, which dominates and soaks up living labor-power" (548). Here we see the completion of capital's domination of the worker, one based on the initial fundamental separation of "mental" and "manual" labor.

> Factory work exhausts the nervous system to the uttermost; at the same time, it does away with the many-sided play of the muscles, and confiscates every

> atom of freedom, both in bodily and in intellectual activity. Even the lightening of labor becomes an instrument of torture, since the machine does not free the worker from the work, but rather deprives the work itself of all content. (548)

Marx argues here that easier work becomes painful "torture" for the worker, as it doesn't free the worker from work but rather renders work meaningless—in other words, it transforms it into empty, menial labor. Work deprived of content is work that offers no pleasure or satisfaction in its doing, no sense of purpose or accomplishment; it's work that is completely alienating. Such "contentless" work can be performed by anyone, meaning workers can be replaced at any time, further weakening the position of labor in the workplace.

Marx calls this state of affairs in which labor can only be performed in conformity to the will of management—that is, under the supervision and command of capitalists—the "tyranny of capital."

Once the tyranny of capital has been achieved, the dramatic rises in productivity that result appear as the sole possession of the capitalist. While in reality it's a form of "cooperative labor" that has led to these increases in productive force, these increases appear not as increases in the productive force of *labor*, but rather as increases in the productive force of *capital.* Workers aren't working cooperatively according to their wishes, but rather are being coerced into it by outside forces.

Workers can make a car or a computer because capital has gathered them together to work together according to its wishes. We thus find ourselves unable to make anything at all without the supervision and command of capital. This is why Marx asserts that the expansion of productive force paradoxically only strengthens the domination of capital over our lives.

POWERLESS PRODUCERS ARE POWERLESS CONSUMERS

As workers, we've become powerless, unable to make anything on our own, and that casts a large shadow over our day-to-day life. At the same time, we find ourselves becoming powerless consumers as well.

We previously discussed the internalization of capitalistic "efficiency" in the form of "cost-performance thinking." Under capitalism, we think of time as something that should only be used to maximize profits; other activities are seen as inefficient and wasteful tasks that should be contracted to others. Capitalists see opportunity in this relationship, and services and products are invented that perform tasks deemed to be a waste of time.

Who is using these services? People who've achieved a certain amount of financial and professional success, surely. But how independent are they—how much agency do they really exercise over their lives? Not as much as it might seem. After all, they cannot cook for themselves, or keep their own houses. They've become deeply dependent on the care and service work of others.

Consumers become powerless to do anything other than choose from among the commodities and services offered by the marketplace. It's true that as long as we have enough money, we can look at the menus offered to us and order whatever we want. But if it's not on the menu, we can't get it. In that sense, we're disempowered and captive to the market.

For example, if you go to the supermarket, the fish available there are the ones already cut into fillets or sashimi. No matter when you go to the fish counter, the same kinds and cuts of fish will be for sale. This limit becomes the limit of what ends up on your dinner table.

I was living as an exchange student in Germany when one day, an old toilet in my apartment backed up. I tried using the plunger, but to no avail. My first thought was to call someone to fix it for me. But I couldn't think of the right German words to look up and find such a person. I asked a fellow student, a German, for advice, who, after replying calmly that these things happened all the time, opened up the drainpipe and fixed it for me himself.

Germany is still a capitalist society, but it's also a place where I frequently encountered situations in which people were empowered to make or do things themselves rather than contract the task out to someone else. It's a relatively "eco-minded" society where people tend to repair old cars and appliances rather than replace them. In this sense, Japanese and US consumers' overreliance on costly services and their propensity to throw away household appliances as soon as something malfunctions make these more "capitalistic" societies.

PUSHED OUT OF EMPLOYMENT BY INCREASED PRODUCTIVITY

There's another reason why an increase in productive force is not always an occasion to rejoice: the problem of relative surplus population.

Let's say mechanization allows a factory's productivity to double. This means that the number of workers necessary to make the same quantity of a commodity will decrease by half. If the economy is good, this isn't much of a problem, as the capitalist will then double the scale of production as well. But there's a limit to how much demand a society can support, and industry will inevitably run up against it. That's when capital no longer needs workers. This is especially true in economies like Japan's, which has experienced steady decline over time. If productive force continues to rise, industry has no choice but to ruthlessly restructure and lay off workers to cut costs.

Once this happens, the labor market will overflow with workers that are "surplus" relative to the demands of capital, resulting in a "buyer's market" for labor-power.

> The capitalistic application of machinery on the one hand supplies new and powerful incentives for an unbounded prolongation of the working day, and produces such a revolution in the mode of labor as well as the character of the social working organism that it is able to break all resistance to this tendency. But on the other hand, partly by placing at the capitalists' disposal new strata of the working class previously inaccessible to them, partly by setting free the workers it supplants, machinery produces a surplus working population, which is compelled to submit to the dictates of capital. (531–2)

Because machinery now performs physically arduous tasks, not only untrained apprentices but also women and children can begin working as part of the "new strata of the working class previously inaccessible" to capital. The mechanization of agriculture through the introduction of tractors and harvesters creates a surplus population in farming villages, which leads many young people to move into cities seeking work. Relative surplus population goes up as these parts of the population join the labor market.

As more people appear in cities willing to work for low wages, perform arduous work, and simply take a job, any job, no matter how demeaning, those already working in the factory push themselves to work even harder and longer, fearful of losing their jobs to those more desperate.

Ironically enough, though, the harder that workers work, the more productivity increases, which leads the capitalist to think, "Well, if they can work this hard, I don't need a hundred of them to maintain production levels—eighty will do!" This increases relative surplus population even more.

> The over-work of the employed part of the working class swells the ranks of its reserve, while, conversely, the greater pressure that the reserve by its competition exerts on the employed workers forces them to submit to over-work and subjects them to the dictates of capital. (789)

This situation divides the employed from the unemployed, and workers unable to unionize end up in an even weaker position vis-à-vis capital. Disempowered and unable to resist, workers find themselves accepting "over-work." Marx addresses the problem located at the heart of this vicious cycle:

> Within the capitalist system all methods for raising the social productivity of labor are put into effect at the cost of the individual worker; that all means for the development of production undergo a dialectical inversion so that they become means of domination and exploitation of the producers; they distort the worker into a fragment of a man, they degrade him to the level of an appendage of a machine, they destroy

> the actual content of his labor by turning it into a torment; they alienate from him the intellectual potentialities of the labor process in the same proportion that science is incorporated in it as an independent power; they deform the conditions under which he works, subject him during the labor process to a despotism the more hateful for its meanness; they transform his life-time into working-time, and drag his wife and child beneath the wheels of the juggernaut of capital. (799)

What is a juggernaut? We think of it as an unstoppable, merciless force of absolute destruction. The word comes from the name of the Hindu god Jagannath, an avatar of Vishnu, who is honored by enormous temple cars. At the time Marx was writing, Europeans were being fed tales of true believers throwing themselves under the wheels of these temple cars in the midst of religious frenzy. This is the image Marx uses to characterize the cruelty of capitalism.

As the constant churn of M–C–M^1 powering capital's movement speeds up and grows ever stronger, it swallows all parts of society whole, annihilating the lives of workers along the way. Yet workers submit to this annihilation—or rather, in order to live, they willingly throw themselves and their families beneath its wheels in the form of self-imposed regimes of excessive work.

THE LIE KNOWN AS "MANAGEMENT PERSPECTIVE"

The juggernaut of capital, which demands extreme sacrifices from workers, has only increased in scale as it has rolled into the present day. Capital's separation of conception and execution has only grown more stark as well.

As we saw, Taylor developed his management system to force workers into "voluntarily" maximizing the effort they put into their labor, so that their work becomes as efficient as possible. But today's workers are expected not just to work hard, but also to keep a "management perspective" in mind, which is meant to encourage them to work according to the wishes of their internalized boss. To that end, employees are often asked to study a company's handbook and "management philosophy."

At first blush it may seem as though discourse on "management perspective" allows workers to access conception and not just execution. But this is not the case—rather, it's a way to con workers with agreeable language that some scholars have mistaken for a sign we've overcome oppressive management systems like Taylorism that robbed workers of their agency. However, this discourse does nothing to help the worker overcome the separation of conception and execution. Rather, it represents the further internalization of what managers and supervisors already want, but now in the form of an instruction manual urging workers to become even more obedient to the wishes of those already monopolizing conception.

Conception refers to the ability to judge for yourself how best to adapt your work to specific situations. This is not someone working at a store selling fast fashion looking at the local weather conditions and deciding whether to dress a mannequin in a T-shirt or a dress shirt. Conception is a more expansive thing, encompassing how employees deal with the public, how they decide what items to stock and sell, or how long they decide to leave the store open.

In this sense, workers who lack the power to make these types of decisions but are nonetheless urged to work with the "management perspective" in mind are simply being forced to execute tasks while being barred access to conception, even as they're encouraged to think they have complete autonomy. And insofar as workers continue to believe this, alienation appears to have been diminished—but in truth this is a sign that workers have become thoroughly subsumed by capital. It is far more convenient for the capitalist if workers "voluntarily" internalize the capitalist's perspective.

It should now be clear that the promise that technological advances like AI and robotics will enable workers to enjoy more freedoms is an enduring lie.

For example, during the COVID-19 pandemic, I gave working for Uber Eats a try. It had often been referred to as an example of the "sharing economy," in which goods and services could be shared and exchanged while work could be done "freely" whenever one liked, all thanks to smartphone technology. But as it turned out, the only labor involved was delivering food. The real work I ended up performing was staring at the screen of my phone and following the directions until I reached my destination and handed over the order.

In that sense, the content of my labor was dictated by Uber's algorithm and the instructions from my GPS; the entirety of my task was to deliver food before it got cold. Labor cut off from conception has no room for creativity or communication with others. All that mattered was upping the efficiency of my deliveries. And in the end, I found it was incredibly unsettling to have my performance constantly assessed by a machine.

BULLSHIT JOBS

On the other hand, it's clear that Osborne's prediction that machines would steal all jobs from humans is also unlikely to come true. In the same way that food still needs a human to deliver it despite recent technological advancements, many jobs remain that machines still can't do.

For example, employees at convenience stores are responsible for a wide variety of tasks, making them nearly impossible to replace with a single machine. Other jobs, like nursing and daycare, might seem menial at first but are incredibly difficult, essential forms of work that depend on an acute sensitivity to the variable needs of others and would be prohibitively costly to make robots do.

The fact is, many jobs are much *cheaper* when performed by humans, meaning that there is little danger they will be performed by robots anytime soon. Instead, the real problem is that the essential workers who do these jobs—jobs that only humans can do and that are indispensable to the functioning of society—are burdened by long work hours and low pay.

At the same time, jobs that aren't particularly important to the functioning of society, that even the people doing them have a hard time finding meaning in, are comparatively well-compensated. As the cultural anthropologist David Graeber put it, these kinds of "bullshit jobs"—in advertising and consultancy, for example—have proliferated dramatically in the last few decades.[11]

Attending pointless meetings, filling out endless paperwork, writing mindless ad copy, undergoing training in office manners—these are all features of bullshit jobs. They're symptoms of the rise in productive force creating an environment that cannot support forty hours of labor a week from its elites, in turn creating meaningless tasks to fill the work time. Graeber points out that the reason Keynes's prediction didn't ultimately come true is because he failed to see how capitalism can and will create a huge volume of trivial and inconsequential jobs to keep everyone working.

Even as these kinds of highly paid, yet aimless, bullshit jobs proliferate, the essential workers performing the jobs that are truly foundational to society are forced to endure increasingly miserable labor conditions. This is the state of contemporary society as capitalism reaches full maturity. Both unfulfilling, meaningless labor and exhausting, overlong labor immiserate workers in the end.

In other words, despite the huge wage disparities among them, the vast majority of workers in today's society are alienated from their labor.

RECLAIM AUTONOMY!

Yet this doesn't mean that the solution is to create a society in which humankind is freed of labor entirely, living lives devoid of any work at all. Marx didn't envision a future society in which we all sit idly by drinking beer while machines and robots do all our work for us.

To reiterate an earlier point, Marx asserts that the real problem is the separation of conception and execution under the domination of capital, which robs labor of its content. The way to restore richness to the wealth represented by human labor is to heal this divide between conception and execution and restore autonomy to the worker. It's not just a matter of freeing workers from cruel overwork—we must also remake labor into something compelling and fulfilling. In other words, Marx would argue that getting rid of labor entirely through the use of robots and AI would be to locate the root of the problem in the wrong place.

Workers in the future society envisioned by Marx are "fully developed individuals"—not atomized workers making money mindlessly turning screws, but rather workers who are skilled in both the conception and execution of their tasks and are able to marshal the wealth of their own labor-power to enrich the wealth of society overall. In this way, we could support each other while still exercising our autonomy, refining our skills, and experiencing satisfaction and passion. This is how, according to Marx, we can finally heal our alienation.

TOWARD THE OVERCOMING OF ALIENATION

Taking into account the proliferation of bullshit jobs in today's society, we might take this observation a step further: Isn't the actual innovation we need to restore richness to our lives the overcoming of alienation?

We imagine innovation as something that makes our lives better, or at least more convenient. But the commodities around us—our refrigerators, televisions, washing machines—have remained essentially the same for the past fifty or more years. Even our ever-updating smartphones no longer change much between iterations. Instead, all that's happening is that our refrigerators, microwaves, and smartphones keep acquiring superfluous functions. Even those whose job it is to develop these new functions must sense their uselessness.

One of the key causes of stagnating innovation is the extreme division of labor and the concurrent separation of conception from execution. Capitalism wrings every last drop of intellectual innovation to keep pace with the demand for new commodities, but it does so at a remove from the input of actual producers or even users, leading to a scarcity of new ideas and the reduction of innovation to minute modifications to existing products. In order to justify this, each infinitesimally minor change is trumpeted as a revolutionary discovery, as industries rely more and more on sectors like advertising and consultancy—prime generators of bullshit jobs—in their efforts to overcome the innovation slump.

The kinds of innovation that cause true paradigm shifts in society will only come once workers are liberated from market competition that only rewards short-term results and performance. Only then will they be able to experiment freely with various ideas in an environment that fosters genuine challenges to the status quo. It's often claimed that the equality brought about by socialism would stifle innovation, but it seems very possible that the opposite is the case.

THE EFFORT TO SAVE SCHOOL LUNCH

In this final section of the chapter, I'd like to discuss efforts to save school lunches in Japan as an example of how we might concretely reunite conception and execution to return autonomy to labor and deepen social prosperity. This example comes from a book by the historian of agriculture Tatsushi Fujihara, titled *The History of School Lunch.*[12]

School lunches in Japan are said to have begun in a private elementary school in what was then the Tsuruoka district of Nishitagawa—now Tsuruoka City—in Yamagata prefecture in 1889. The practice was started not just as a way to provide children and communities with nutritious meals, but as an anti-poverty measure. Lunches brought from home can't help but reflect the economic conditions of those who bring them. The free school lunches not only remedied malnutrition but also promoted an image of class equality. In this sense, school lunches were part of a system for achieving equality in learning.

After the end of World War II, school lunch began to be put to other political uses. The US saw the school lunch system in occupied Japan as a convenient opportunity to sell off domestic surpluses of wheat and powdered skim milk. Furthermore, if students given bread at school developed a taste for it, this would translate into higher sustained sales of flour in the future. This could be said to be a type of subsumption by capital. And in fact, postwar sales of rice went down as Japanese consumers developed a taste for bread and pasta.

After 1960, subsumption by capital enabled the separation of conception and execution within the school lunch system. Before, every school had its own cafeteria and kitchen, as well as its own specialized staff who made hot food every day for students. This was known as the "on-site" system for school lunch preparation.

In 1961, though, the School System Evaluation Committee determined that "a centralized system would be more efficient for providing school lunches to every school," and school lunch centers were set up to provide lunches for surrounding schools in the name of efficiency and cost cutting. This "centralized" system spread quickly throughout the country; while there were 32 such centers in 1963, by 1975 the number had risen to 2,039.

The meals produced by these mechanized centers were necessarily limited in terms of their cooking methods, which meant that the menus were, too; it also meant that meals would arrive from the center to the school completely cold. The introduction of frozen and processed food in the name of streamlining the cooking process caused the nutritional value of the meals to go down and the number of preservatives in them to go up. As soon as efficiency became the top priority, the taste of the meals suffered as well.

There was also increased risk of food poisoning, as a food safety problem at one center could end up affecting all of the schools it served. As recently as 2020, fifteen schools in the city of Yashio in Saitama prefecture were affected in a food poisoning incident that sickened 3,453 adults and children—nearly half the total student and staff population of these schools. This was an obvious result of the "centralized" school lunch system implemented in the area.

The centralized system prioritizes efficiency above all else, robbing cafeterias in individual schools of their access to conception while also depriving them of meaningful access to execution in the form of actual meal preparation—instead, their function is reduced to handing out meals delivered from elsewhere. The result is the reduction of forms of use-value like taste and food safety.

But there is resistance to this wave of centralization, including examples of schools going back to on-site school lunch preparation in an attempt to reclaim food autonomy. Some schools that have returned to on-site meal preparation have set up independent committees to oversee the content of the lunches. Some schools seek not only to provide better nutritional value and taste, but also to incorporate more local products as they diversify their offerings, working with local farmers' associations to enrich children's diets with regional foodways, simultaneously raising their awareness of proper nutrition and their local cultural heritage.

By prioritizing locally available and optimally nutritious ingredients, schools can make seasonal menus, which demand increased know-how and flexibility from the chefs and kitchen workers on-site—yet these workers also enjoy increased satisfaction and fulfillment in their labor, as they now get to witness children enjoying food they prepared for them.

The stark contrast between, on the one hand, the invaluable benefits that were lost when the capitalistic reform of the centralized school lunch system introduced mechanization and menial labor to the meal production and distribution process, and on the other, the richness of the food produced when the system went back to on-site production, provides us with a clear vision of the path we must choose as we create our future society.

The task before us now is, as the young Marx pointed out, to restore human feeling to our lives, which will allow us to appreciate the full richness of both the natural world and our own humanity. We must overcome the tyranny of capital and the alienation of labor, and truly democratize work in a way that will give autonomy and fulfillment back to the worker.

But to do that, it's not enough just to overcome alienation in the realm of human labor. We must also overcome our alienation from nature.

4

GREEN CAPITALISM

A Fairy Tale

CAPITAL LOOTS NATURE, TOO

In the past two chapters, we've looked at how human labor is distorted and disordered under capitalism. In this chapter, we're shifting focus to another major issue with modern capitalism: its effect on the environment. To do this, we will first need to revisit Marx's discussion of metabolism.

Human life is in constant negotiation with nature; human labor is defined as the "acting upon" of the natural world. Because capitalism demands that labor be conducted with the purpose of increasing value, human labor-power ends up abused in the process.

The disordering of labor under capitalism affects more than just the human side of the equation, though—it also affects nature and the environment, giving way to serious and long-lasting problems. As Marx points out, capital "asks no questions about the length of life of labor-power," (376) and we can easily extend this formulation to the natural world:

Capital is not concerned with sustaining the health of the environment.

Despite this, Marx has been frequently criticized as being excessively optimistic about technological development, supposedly ignoring environmental issues out of the belief that future technologies will solve them for us. But looking at his writing on metabolic exchange, it's clear that he in fact did factor humanity's relationship with nature into his analysis of capitalism.

New materials showing the ecological aspects of Marx's thought have recently resurfaced. This chapter will survey this new research to more closely examine the environmental problems caused by the advance of capitalism.

THE ABSURD INEQUALITY OF THE CAPITALOCENE

The worldwide environmental destruction brought about by the explosive advance of global capitalism has plunged the planet into crisis. The amount of fossil fuels consumed by the world population after 1990 now equals half that consumed during the entirety of human history up to that point. This massive consumption of fossil fuels is causing climate change. Related disasters are occurring everywhere: major flooding in Pakistan, wildfires in California, droughts in Africa, rising ocean levels caused by melting glaciers and ice sheets. The effects of climate change have become unignorable even in Japan, which has seen increasingly heavy rainy seasons and intense typhoons, among other natural disasters.

Meat-centric diets have led to extreme deforestation and the worldwide loss of biodiversity. Many scientists believe that these phenomena are partially to blame for novel viruses emerging and causing pandemics among humans.

Environmental destruction has now reached the point that it can never be fully reversed. Many scholars, in emphasizing how capitalism has fundamentally changed the workings of the planetary environment, have begun using terms drawn from geology to characterize the present era: as the Anthropocene—an age defined by human intervention into the natural world—or even, more specifically, the *Capital*ocene.

As you might expect, the Capitalocene is characterized by extreme inequality. Capitalism's rapacious use of the planet has run up a tab to be paid by both future generations and the poor currently living in the Global South. The superrich elite traveling the world in private jets and luxury yachts are responsible for a huge share of the current crisis, yet they refuse to take any responsibility. Perhaps they think they'll be exempt from its effects.

One must ask: Will we survive if we leave the future of humanity up to the forces of capitalism? Will capitalism really generate advances in technology that will save us from catastrophe?

"APRÈS MOI, LE DÉLUGE"

The origins of Marx's ecological turn can be found in the work of his contemporary, the chemist Justus von Liebig, who taught at the University of Munich and published the book *Organic Chemistry in its Application to Agriculture and Physiology* (better known as *Agricultural Chemistry*), whose seventh edition came out in 1862.

Liebig was a harsh critic of then-current farming practices, which quickly depleted the soil in pursuit of high yields (and profits), calling this failure to replenish the land a "robbery system." He went on to argue that if such practices continued, they would destroy the very basis on which the nation was built, leading to the eventual collapse of civilization. This famed chemist had previously extolled the virtues of fertilizer as the answer to soil exhaustion, so his sudden reversal on the issue left a lasting impression on his contemporaries.

Marx was no exception. Liebig's discourse on modern agriculture's robbery of the soil was of particular interest to Marx, and he produced a set of meticulous research notes on *Agricultural Chemistry*. Later, in *Capital*, Marx warns that soil depletion caused by the heedless exploitation of nature would threaten the physical foundation of society itself.

> The same blind desire for profit that in the one case exhausted the soil had in the other case seized hold of the vital force of the nation at its roots. (348)

Marx's excavation into capital's propensity to loot resources caused him to see that soil depletion and the exhaustion of human vitality shared the same origin: the "blind desire" of capital for endless growth.

Moreover, in chapter 15 of *Capital, Volume One*, "Machinery and Large-Scale Industry," Marx discusses the large-scale landholdings supporting the capitalistic management of agriculture. He points out that this looting by capital "disturbs" the metabolic relation connecting humans to nature while also exacerbating the divide between rural and urban workers:

> Capitalist production collects the population together into great centers, and causes the urban population to achieve an ever-growing preponderance. This has two results. On the one hand it concentrates the historical motive power of society; on the other hand, it disturbs the metabolic interaction between man and the earth, i.e. it prevents the return to the soil of its constituent elements consumed by man in the form of food and clothing; hence it hinders the operation of the eternal natural condition for the lasting fertility of the soil. Thus it destroys at the same time the physical health of the urban worker, and the intellectual life of the rural worker. (637)

Agricultural production conducted by capital with the aim of making the most money in the least amount of time leads to heedless monocrop farming (the cultivation of the same crops in the same soil year after year). It takes time, though, for soil to replenish the nutrients consumed by crops. Capitalism might accelerate, but natural cycles remain constant. A rift is thus exposed between the temporality of capital and that of nature, the end result of which, in this case, is soil exhaustion. The land's fertility withers, and harvests dwindle.

Marx also discussed forestry as another example illustrating the same process. It takes decades to grow trees, but capitalism can't wait that long for forests to renew themselves. Instead, capitalists cut down entire forests in search of profit with no concern for natural cycles of regeneration. As technology advances, the pace of deforestation only quickens, resulting in the penetration and destruction of even the most remote wildernesses. Accordingly, the wastelands left behind capitalism grow only more vast.

Capital takes no responsibility for the exhaustion of the soil or the destruction of the forests. Unable to sync itself with the temporality of nature, capital simply takes what it can to transform into commodities and sell with no concern for the future. *Après moi, le déluge!*—this is the slogan of the capitalist. While urban lifestyles become richer, as they are ground zero for commodity consumption, the price of this wealth is paid by the rural areas that suffer from issues like soil depletion and the destruction of the landscape. As a result, the quality of life of those that work and live in these rural areas deteriorates.

But as Marx notes in the previous quote, even workers in the urban centers "blessed" with commodities are forced by capital to toil for long hours at the expense of their physical health and well-being. Ultimately, Marx clarifies, capitalism diminishes societal wealth for everyone.

THE IRREPARABLE RIFT

Marx studied and wrote about soil exhaustion in the nineteenth century, but the looting of nature continues to this day, albeit in slightly different forms: from the extraction of fossil fuels and rare metals from the earth to the destruction of the Amazon rainforests.

Just as life in industrial urban centers in the nineteenth century caused hardship in rural villages, the profligate lifestyles enjoyed by those of us in the Global North are underwritten by the immiseration of those in the Global South. *Après moi, le déluge*—this attitude becomes even more explicit as we pass the environmental costs of our present actions on to future generations. This is what is known as "externalization."

In Brazil, both the Amazon rainforest and the vast grassland known as the Cerrado have been rapidly transformed through clearing and logging operations into pastures and cropland. Why? Because of the production of beef for export to the Global North—the land is cleared to both raise cows and grow the crops to feed them. To take another example, the production of smartphones and computers uses large quantities of rare metals found in South America and Africa. The mining of these metals frequently involves human rights violations, including the use of enslaved labor and child labor—the ILO (International Labor Organization) estimates that over 50 million people are currently working in conditions of enslavement.

Externalization occurs in Japan as well. Japan is the second-biggest importer of lumber in the world, bringing in considerable quantities of wood from Southeast Asian countries like the Philippines, Malaysia, and Indonesia, as well as from Siberia. Paradoxically, almost 70 percent of Japan is covered in rich forestland, providing enough resources that local cedars and cypresses are left unharvested and mountains covered in wild forests, yet large amounts of cheap lumber from overseas are imported anyway, leading to the decline of the domestic forestry industry. In this way, simple cheapness becomes the rationale for pumping yet more carbon emissions into the atmosphere to import lumber produced through the destruction of overseas forests.

All of this is ultimately unsustainable. It reflects how the metabolic relationship between humans and nature has become so distorted on the global scale. Marx predicted this outcome, writing in a draft version of *Capital*'s third volume about this metabolic rift:

> In this way, [large-scale landownership] produces conditions that provoke an irreparable rift in the interdependent process between social metabolism and natural metabolism, resulting in the squandering of the

> vitality of the soil, and trade carries this devastation far beyond the bounds of a single country (Liebig).[13]

The endless movement of capitalism commodifies the entire world while allowing only a small subsection of the population in a small subsection of the planet to profit—this is the monopoly form (or, as in the previous passage, "large-scale landownership"). Globalization has expanded the rural-urban divide far beyond the boundaries of the nation. However, while capitalism seeks the limitless increase of value, the planet itself is limited. Capital always works to externalize the costs of its operations, but the planet acts as a limit on this "externality."

After the collapse of the Soviet Union, the globalization of capitalism sped up even more, globalizing the environmental crisis as well; now we can say with confidence that there is nowhere on earth the crisis has not touched. Even countries like Japan and the United States, well ensconced in the Global North, aren't exempt from the effects of climate change, as super-typhoons and extreme heat have become routine occurrences. The "metabolic rift" has deepened to the point of no return.

THE COLLAPSE OF COMPLEXITY

The metabolic relationship between humans and nature was once a cycle—a process that could be represented as a circle. But the movement of capital is one-way. It unilaterally extracts what it can from workers and the natural world, while foisting the costs of this extraction onto a perceived "outside" in a similarly one-way fashion.

By contrast, ecological systems are complexly interconnected, meaning that when one element is disturbed, all the other components linked to it are affected, too, spreading the disturbance throughout the system. Capitalism is unconcerned with such complexity and mutual interdependence. Money empowers and encourages us to assign a price to all things, making them directly comparable and exchangeable with one another. There's certainly a convenience to this, but it's a simplification that can have extremely violent effects, as it allows things that were originally incomparable—like forms of societal wealth and use-value—to be measured against one another according to the abstract concept of "value."

The mutual interdependence of an ecosystem and the valorization of capital operate according to two completely different logics, and it's impossible to reconcile the great divide between them. The simplifying logic of capital, which seeks to commodify anything and everything and which operates solely on the basis of the "value" of the commodified object, cannot ever truly grasp the complexity of the process by which an ecosystem—a system that includes humankind!—operates.

The same holds true for us as individuals. We're so preoccupied with cost-performance thinking, diligently weighing the cost of a steak against the cost of a computer, that we lose sight of the *true* cost of the things we use to make our lives.

What went into producing that steak? How much carbon dioxide was emitted to feed the cow it came from? To deliver its flesh to us? We simply don't think about it.

It is precisely this sort of simplification that is necessary for capital to increase itself so swiftly and completely, opening up a rift in our metabolic relation to nature and in turn causing the current environmental crisis. We can no longer afford to keep the true cost of commodification and the valorization of capital out of sight and out of mind.

Don't misunderstand me—Marx, too, recognized the fact that capitalism had, in its fashion, enriched our lives as well. However, he also saw that its fixation on raising profits by recklessly increasing productivity by any means necessary was also a major hindrance on humanity's ability to protect collective planetary abundance for future generations.

Marx speaks to this in a famous passage from the preface of his 1859 treatise, *A Contribution to the Critique of Political Economy*:

> At a certain stage of development, the material productive forces of society come into conflict with the existing relations of production or—this merely expresses the same thing in legal terms—with the property relations within the framework of which they have operated hitherto. From forms of development of the productive forces these relations turn into their fetters. Then begins an era of social revolution.[14]

It's impossible to raise productive forces any more under capitalism without completely destroying the planetary environment. We cannot manage the environment—a resource that ultimately belongs to no one—in a sustainable way as long as we operate within a system founded on ceaseless exploitation in the name of private ownership and profit seeking. We've reached the point at which capitalism itself has become the "fetters" on society's development.

We can thus say that Marx advocates for revolutionary change that would usher in a completely different social system, a change that urgently needs to happen before the "irreparable rift" in the metabolic relation between humans and nature destroys civilization for good.

THE SUBSUMPTION OF NATURE NEVER STOPS

The fact remains, though, that even if we address these sorts of "fetters" head-on, capitalism won't stop. Instead, it will just give lip service to superficial solutions addressing the environmental crisis while continuing to commodify nature even further. In the previous chapter, we saw how labor became subsumed by capital and how workers lost their skills. The same sort of process is happening in agriculture as well.

The advent of the twentieth century saw agricultural productivity continue to rise when soil exhaustion was resolved through the usage of chemical fertilizer and pesticides. In Japan, organic farms make up a mere 0.6 percent of the farmland currently in production—fertilizers and other agrichemicals have become a prerequisite for maintaining productivity. Farmers themselves generally don't make their own agrichemicals, so they must purchase them as commodities, thus making the average farmer dependent on the goods and technologies sold by large-scale industry.

If you rely too much on chemical fertilizer to boost your productivity, though, the soil's ability to retain water deteriorates, and the threat of soil exhaustion returns. Overuse of pesticides and herbicides, on the other hand, can make crops prone to disease while also hastening the emergence of so-called "superweeds" resistant to herbicidal control.

In other words, we are caught in a vicious cycle: Overcoming one metabolic rift entails the use of more and more fertilizer and new forms of agrichemicals that ultimately open up an entirely new metabolic rift. Yet to capital, this represents nothing more than another exciting business opportunity.

The same goes for the use of hormones and antibiotics in the livestock industry. The production process has been continuously transformed to suit capital's interests, encouraging faster growth and the extraction of more meat per animal. The overdevelopment of factory farming has resulted in worsening living conditions for the animals, as well as increases in complications like water pollution, the spread of disease, and the degradation of the natural landscape.

But even as these problems arise, capitalism marches on. The pursuit of ever-higher levels of productivity have led in recent times to escalated interventions into the workings of nature itself. Biotechnologies like genetic recombination, genome editing, and lab-grown meat are playing an increasingly large role in the agricultural and livestock industries.

Yet we must ask ourselves if it's wise to adopt new technologies with the power to transform nature in the name of raising profits. Only large-scale industry has the means to develop and implement these new technologies, and this may well lead to a future in which basic needs like food are entirely monopolized by a select few corporations.

As the Japanese population has been getting steadily older, the number of farmers in the country has dropped dramatically. In 2000, the number of agricultural workers in Japan was around 2.4 million, while today that number has fallen to about 1.4 million. The industrialization of agriculture is spurring this downward trend even further.

"Smart farming" techniques that automate and mechanize agriculture—for example, through the use of drones—will only grow more prevalent as time goes on. As more sensors, cameras, and other management systems are put in place, the monitoring and management of crops and livestock will become progressively more automated.

The amassing of data culled from devices that make up the "internet of things" gives way to the quantification of the knowledge, judgment, and skills needed to maximize harvests. This, in turn, enables the construction of management systems for livestock and agriculture that are standardized and duplicable without specialized knowledge or experience, which means that these jobs are gradually being given to inexperienced student workers or foreign "trainees," leading to the de-skilling of the industry as a whole. This is the same process as that described in the previous chapter's discussion of Taylorism and its insidious effect on factory work. It's no coincidence that the largest private owner of farmland in the US is none other than the founder of Microsoft, Bill Gates.

The introduction and development of smart farming technology necessitates this scale of capital investment. As all agriculture becomes concentrated into large-scale industrial farming, small-scale farmers will end up being pushed out of the industry, along with the traditional knowledge and experience they hold.

The industrialization of agriculture may boost productivity greatly, but even if profitability goes up, whether this is actually sustainable becomes a whole new problem. If North and South America continue to move in this direction, more likely than not the spread of profit-forward agribusiness practices will lead to the concurrent widening of gaps within the industry, and deepening of the rift in the metabolic relationship between humans and nature.

THE TURN TOWARD ECOLOGY

Once we begin to see the issue this way, it becomes easier to understand why Marx studied the natural sciences so diligently. Liebig's analysis of agricultural techniques was just the gateway to further investigations of botany, geology, and mineralogy; Marx was an impressively comprehensive and pluralistic researcher.

Despite this, Marx's investigations into the natural sciences have long been ignored until only very recently. Renewed interest in Marx's work during this period is largely due to an international group of scholars contributing to the ongoing MEGA project—the *Marx-Engels-Gesamtausgabe* (in English: *The Complete Works of Marx and Engels*).

The MEGA project was started in the Soviet Union and East Germany, but after 1990, many scholars from all over the world joined, and its scale has expanded to a projected one hundred–plus volumes. The materials gathered include not only Marx and Engels's previously published works, but also prepublication drafts and manuscripts, as well as newspaper articles, correspondence, and notebooks. But perhaps the most important materials are Marx's vast research notes, which are being published in MEGA's fourth section.

Marx meticulously wrote out extensive quotations from his readings. Marx himself wrote, in a letter to his father dated from when he was nineteen years old and attending the University of Berlin, that he had "made a habit of the practice of excerpting passages from out of all the books that [he] read."[15] He kept this habit throughout his life.

I personally oversaw the editing of the eighteenth volume of the fourth section of MEGA, which covers the material from the notebooks relating to his research into the natural sciences. These notes make it clear that in addition to the problem of soil depletion, Marx was interested in the overlogging of forestland, the exhaustion of coal resources, and the negative impacts of selective breeding on the health of animals. Regarding the latter, he expressed revulsion at the idea of breeding qualities into sheep that only benefited humans and ultimately jeopardized the health of the sheep themselves, saying it "nauseated" him.[16]

This was, in a sense, a shocking revelation. After all, Marx's philosophy has been frequently criticized as lacking an interest in the value of nature and possessing no thought or theoretical dimension regarding the environment. He's seen by many as a productivist: someone who thought that if productivity were only raised high enough, large-scale social issues like the environmental crisis and poverty would be solved.

His research notebooks, though, reveal the opposite: Marx was in fact paying close attention to the ways in which capitalism was deepening the exploitation of nature. It's with this insight in mind that he wrote the passage previously quoted from the manuscript of *Capital*'s third volume, which states that the rift capitalism has opened up in the metabolic relation between humans and nature is "irreparable."

The notebooks dating from after *Capital*'s publication in 1868 reveal that Marx was greatly influenced by the German agriculturalist Karl Fraas, who argued that overlogging had caused changes in the climate that resulted in the collapse of both indigenous agricultural practices and, eventually, their civilizations. Marx went so far as to assert that he saw "an unconscious socialist tendency" in Fraas's criticism of the exploitation of nature and his call for more sustainable methods of managing forestland.[17] The research that scholars like Liebig and Fraas did on sustainability contains within it a demand to restrain capitalism's drive for endlessly multiplying value and replace this system with a more rational, planned system of economic production.

Stimulated by these contemporary debates within the natural sciences late in his life, Marx began to imagine a postcapitalist world in a way that was inseparable from the problem of environmental sustainability. This is what is known as ecosocialism—the wish not only to establish economic equality among humankind, but also to develop a more coherent and generous approach to managing the metabolic relation between humans and nature. These days, in an age defined by the global environmental crises spurred by capitalism, ecosocialism is being trumpeted once again.

MARX'S LATE-PERIOD THOUGHT THAT DIDN'T MAKE IT INTO *CAPITAL*

> Freedom, in this sphere, can consist only in this, that socialized man, the associated producers, govern the human metabolism with nature in a rational way, bringing it under their collective control instead of being dominated by it as a blind power; accomplishing it with

> the least expenditure of energy and in conditions most worthy and appropriate for their human nature.[18]

This is from the seventh part of the third volume of *Capital*, part of a section recovered by Engels as fragments within the manuscript for what became part six in the present version. Marx understood that the new society replacing the present capitalistic one should consist of "associated" workers explicitly managing the metabolic relation between humankind and nature in a more considered and inclusive way—that is, in a way that values long-term sustainability over short-term overexploitation.

To *associate*, in this context, means to voluntarily come together to work toward a common purpose; that is, to *cooperate*. We'll look at this concept in more detail in the next chapter, but for now it's important to understand that this concept includes an ecosocialist dimension.

But what exactly can we do to usher in this new society, to heal the rift in the human metabolism with nature? The fact is, there are no concrete steps to be found in these pages. I vividly remember reading *Capital* during the winter vacation of my first year at university and feeling terribly disappointed to reach the end and realize there was no real description of the postcapitalist society to come.

Why not? One reason is that *Capital* is incomplete. The first volume was published in 1867, and then Marx plunged back into his research and writing. Fifteen years passed, during which he completed the second volume through its eighth chapter, but then, before he could complete the rest, he passed away.

The second volume had been revised seven times by then (at least through its eighth chapter), and it was quite close to completion. The surviving draft versions of what was meant to be in the third volume, however, were all written between 1864 and 1865, before the publication of the first volume. After the initial publication of the first volume, a second edition and a French edition were published five years later. These included many revisions that incorporated some of the new research Marx had embarked on during that time, but this only scratches the surface of the later developments in his thought.

The version of the third volume of *Capital* available to us was developed posthumously through the hard work of Marx's lifelong friend and collaborator Friedrich Engels. The version we read today that appears as a seemingly complete text was shaped by Engels to read that way—and of a piece with the other two volumes. Engels sanded down any rough edges or points of contention that remained unresolved in the original.

There's no question that Engels performed a great service by assembling and publishing *Capital*'s third volume after Marx's death. If it weren't for his immense efforts, *Capital* would surely not have gone on to be so widely read and studied; after all, the first volume had only sold around a thousand copies at that point. Without Engels, Marxist thought wouldn't have spread throughout the world the way it did.

Yet at the same time, the more Engels worked to turn Marxism into a coherent system of thought, the more he ended up obscuring the unresolved points that Marx was still grappling with, making it harder to see the evolution of Marx's thought near the end of his life. These new insights frequently forced Marx to reconsider much of his own previous thought, making it impossible to incorporate into the existing drafts of *Capital*.

So what are these ideas that couldn't be included in *Capital*? And what is Marx's real vision of the future society that could repair the "irreparable" rift in the metabolic relation between humans and nature—the vision that couldn't be articulated in *Capital* itself? Now, more than ever, we need clear answers to these questions. In other words, we need a theory of communism for the twenty-first century.

Yet there are surely many who, despite being sympathetic to Marx's critique of capitalism, reject the suggestion of a modern communist society by arguing it would simply repeat the mistakes of the Soviet Union. In the next chapter, I will look at the problems with "socialism" as it has existed in the Soviet Union and China and map out how we may avoid repeating the mistakes of the past.

5

GOODBYE LENIN!

TOWARD RESTORING RICHNESS TO SOCIETAL WEALTH

I've tried to use metabolic thinking to explore the pitfalls and dangers of capitalism. In capitalist societies, societal and natural wealth is exploited and ultimately destroyed. The climate crises and wealth gaps that have come to characterize our present moment are symptoms of capitalism's mechanisms.

Now, in the last two chapters of this book, I'd like to shift our focus to how we might restore our shared social and environmental abundance in the face of climate catastrophe and mass inequality and precarity.

Marx envisioned that the full restoration of societal wealth could only take place once we had fully disavowed and abandoned capitalism. We must avoid believing it can occur *within* capitalism, as if it were simply a matter of raising taxes and redistributing the funds or paying workers more.

Marx famously called this new postcapitalist social system that would arrive "communism" or "socialism." (He used the two terms more or less interchangeably.) The ambition and radicalism of this foundational argument is what continues to make Marxist thought so compelling, and also why so many have seen it as so dangerous.

So what sort of society is this? How can communism help bring richness back into our lives?

THE CONSERVATISM OF COST-PERFORMANCE THINKING

The vast majority of people believe that capitalism is a fact of life, incapable of being changed much less abandoned. They argue that criticizing capitalism is a waste of time, and hopelessly inefficient from a cost-performance perspective.

Many people have resigned themselves to capitalism's dominance, and instead tried to find ineffective alternatives or reforms within the capitalist model. Some feel that if they continue to be hardworking and law-abiding citizens, their economic and social positions will remain stable.

The fact is, though, that the future of the companies we work for is hardly guaranteed; we may work diligently straight up to retirement and still find ourselves without a pension waiting for us. Can we ever truly depend on the economy? But it's precisely this deep and pervasive uncertainty about the future that makes us turn inward.

As soon as this uncertainty appears, capitalism is there to exploit it. In this case, it's through the concept of investment. Most of us have at least heard of things like tax-exempt mutual funds (such as NISA in Japan) or pension programs (such as iDeCo in Japan or 401(k) plans in the US), and many of us participate in one or more right now.

A program like NISA is a system for low-level tax-exempt investment; that is, a system by which a set amount of money is invested every year, and the profits generated by this financial product are tax-free. In the NISA program, yearly investments up to $18,000 are exempt from taxation for twenty years. (The "New Form of Capitalism" program, first proposed by former Prime Minister Fumio Kishida, promises to revise this to have no time limit at all.) A program like iDeCo is a self-managed pension system whereby the invested money is, in principle, left untouched until the investor turns sixty; the sum accumulating in the account is taken from one's take-home pay. By putting money in these kinds of programs, an investor can potentially make their initial investment grow more than if it was in a regular savings account, while also reducing the amount of taxes they pay. Both systems were implemented in Japan as a covert response to the report by the Japanese Financial Services Agency that the average person would have to save up at least $200,000 to afford to retire.

In other words, these are times that compel us all to become "investors" in order to survive our later years. The nation is no longer able to guarantee our well-being by providing adequate post-retirement financial support or health care, so these things have become matters of personal responsibility—we are asked to invest ourselves and take on the risks as individuals.

Once you begin to pay into a program like NISA, you've become an investor: preoccupied with following the stock market every day, hoping and praying it will only go up.

This is how we become conservative, personally invested in the continued rise of the stock prices that guarantee our well-being after we retire. In other words, we fervently wish for capitalism's prosperity to last forever. This causes widespread unpopularity of measures to redistribute wealth such as raising income taxes or limiting the number of shares one can hold of one's own company's stocks, while measures meant to raise stock values—such as lowering corporate taxes or deregulating the financial market—gain enthusiastic support.

But these policies end up benefiting large stockholders and big business, not small investors. While we enjoy taking home modest yields of 10,000 yen [$63.42 USD] or so, the superrich enjoy profits of millions, even billions of yen.

Even so, our investment activities lead us to internalize the viewpoint of the capitalist and forget that in truth, we're still merely workers. The economist Makoto Nishibe has written incisively about this phenomenon, a consequence of what he calls "free investment capitalism."[19]

In a free investment capitalist society, we are encouraged to see our every choice and act as a type of investment; our entire society ultimately becomes an exercise in cost-performance thinking. What is the cost-performance ratio of getting married? Of having children? Of culture? Of democracy?

Most of the things we spend our lives doing are, naturally, unconnected to wealth creation. When we allow cost-performance thinking to inform our outlook too thoroughly, though, the majority of what gives meaning and texture to our lives begins to seem like a waste of time: communication, culture, politics . . . the list goes on. This spells the end of community and mutual aid, and the withering of societal wealth overall.

Doctor and media commentator Takeshi Yōrō once joked during a conversation with me that from a cost-performance perspective, the most efficient life plan would be to "hop straight in the coffin."[20] But the fact is, it's a way of seeing life that drains it of meaning, the ultimate step in capitalism's "subsumption of the spirit."

COMMUNISM'S IMAGE PROBLEM

In Marx's day, the subsumption of the spirit by capital had not yet come to pass. But we can see how the current situation in Japan, the United States, and elsewhere is the logical extension of the setbacks of capitalism that Marx diagnosed in his work.

With our spirits subsumed, it becomes harder for us to imagine the society Marx asserts is waiting for us on the other side of capitalism. As the Marxist literary critic Fredric Jameson once put it, "It is easier to imagine the end of the world than to imagine the end of capitalism."[21] Even those on the left have trouble doing so—it's easy to see why most people resign themselves to simply trying to find a way to survive under capitalism as best they can and leave it at that.

Our imaginative capacity has become so impoverished that we've lost the ability to envision a society beyond capitalism. For most readers, when they hear the words "communism" and "socialism," it's likely the only things that come to mind are examples like the Soviet Union, China, Cuba, or North Korea.

And indeed, these are nations where "communist" or "socialist" parties hold or once held sway, using language from Marxist theory to justify their reforms and policies. And many Marxists in Japan supported the Soviet Union when it still existed, and there are some who see present-day China as (at least a step toward) socialist governance. But for most people, these places have only negative associations: dictatorship, executions, famine, and so on.

I take these concerns very seriously, and in the rest of this book I'll offer a critique of Soviet and Chinese-style "socialism," and describe why I'd like to bid a permanent goodbye to Marxism-Leninism in general, as I believe it fundamentally misunderstands Marx's project to create a future society founded on restoring the commons.

COMMUNISM AND THE SOVIET UNION ARE TWO SEPARATE THINGS

Even if the Soviet Union or China are seen as societies that have overcome capitalism's challenges, most people have little interest in living there: Japan or the US surely seem like better choices. I, too, feel the same way.

Yet when people hear me discuss communism, many immediately assume I'm praising such places; they also chalk the collapse of the Soviet Union up to a failure in Marx's thought. When I participated in a debate with Czech economist Tomáš Sedláček on the Japanese television program *Capitalism and Desire 2022: Beyond the Dilemmas of Growth and Distribution*, I was shocked how, despite my best efforts to explain their differences, the communism dreamed of by Marx and the realities of the Soviet Union were continuously conflated during the discussion.

The overwhelming impact of the Soviet system on world history is deeply responsible for the ongoing misconceptions about communism. It's incumbent on any Marxist to forthrightly address how to avoid repeating the tragedy of the Soviet Union. But it's also true that the failure of the Soviet Union doesn't require us to repudiate all Marxist thought, for there are critical differences between the two.

THE LACK OF DEMOCRACY

In the Soviet Union, minimum levels of health care, education, and daycare were available without charge, and there were systems in place meant to provide individuals with basic needs at very low cost—places to live, daily necessities, public transportation, and so on. Workdays were short, and a certain level of equality between the sexes was achieved in pay and career advancement. There was indeed a version of socialism in place, and I would never argue that every part of the Soviet system was bad in that regard.

But the main reason people have a negative impression of Soviet and Chinese forms of socialism is their lack of democracy. In these so called socialist countries, there's only one ruling party—the Communist Party—and this causes serious harm.

It's an irrefutable fact, for example, that under Stalin's dictatorship, millions were imprisoned in forced labor camps known as gulags, while many others who criticized the regime were killed as "traitors." During the massacre known as the Great Purge of the 1930s, even loyalists ended up murdered: not just fellow Communist Party members who opposed Stalin, but scores of soldiers, scholars, cultural figures, and average citizens were executed as "spies" or "counter-revolutionary elements."

The number of victims, according to official statistics, reached upward of eight hundred thousand, but the true number is likely somewhere between eight and ten million.

The lack of democracy under the Communist Party enabled these unimaginable crimes to take place. Without democratic mechanisms in place, there was nothing to check the violence of the state, allowing the Stalinist regime to massacre so many of its own citizens.

Similar tragedies occurred in China as well; for example, the Cultural Revolution or, more recently, the Tiananmen Square massacre. The current persecution of minorities like the Uyghurs or those calling for democratic reforms in Hong Kong are examples of the same repression in a new form. Dictatorship without democracy is not the future society we should be working toward.

WHEN GOVERNMENT OFFICIALS BECOME A SPECIAL CLASS

So-called state socialism lacks democracy, which means that it also lacks freedom of expression and association. As an example, let's look at an incident that took place at Peking University in 2018.

A Marxist student group at the university (called the Peking University Marxist Society) became concerned about the working conditions of laborers in Shenzhen. As Marxists, they expressed solidarity with these workers' pleas for better conditions. The students were ultimately confined to the campus and forced to break up their group. Some in the group even reportedly went missing.[22]

In this existing "socialist" country, expressing solidarity with workers based on Marxist theory has become a very dangerous activity to engage in. What is there left to do but wonder at the irony?

Indeed, who would visit present-day China and feel like they were visiting a socialist country at all? There might not be Amazon, but there is Alibaba. Google and Facebook are inaccessible, but there's Baidu and Weibo. Nike, Starbucks, and McDonald's all have locations there, and if you visit Shenzhen, you can see completely automated high-tech factories where electric cars and precision instruments are manufactured around the clock. You'll feel as though you were visiting a society more capitalistic than even Japan or the US.

Searching for socialist features in present-day China, you might notice that a much higher proportion of banks, landholdings, and corporations are nationalized compared to your average capitalist country, something China has in common with the old Soviet Union.

Because of this, the common understanding of socialism defines it in terms of ownership: capitalism as private ownership, socialism as nationalized ownership. The assumption is that the exploitation of workers can only be abolished through a planned economy, which entails widespread nationalization.

The fact is, though, that it is limiting and problematic to define socialism this way. Critically, the nationalization of the mode of production does not itself liberate workers. Who ends up planning a society's overall production when it's nationalized? The workers? The answer is no—the task falls to the Party and its officials, who have jurisdiction across the entire country. These officials take the place of capitalists as those who hold the right to decide how production will proceed, meaning that workers must now work under their direction.

From the workers' point of view, the forces at the top may no longer be capitalists but government officials—they now work for a nationalized enterprise, not a private one—but in terms of having to obey orders handed down from high in the workplace, there's not much difference at all.

As nationalized industry grows, the power wielded by officials grows, too. They frequently drive workers into overwork, using the results to expand their investments into new areas, all in the name of competing to win out over Western-style capitalism and replace it with socialism. In this way, government officials become a special class within society.

As domination by the Party and its officials becomes increasingly uncontestable, the position of workers and the common people weakens, and democracy itself is threatened. It's no coincidence that attempts to establish a planned economy by nationalizing the mode of production have so often resulted in dictatorship.

THE SOVIET UNION AND ITS COLLAPSE

All existing "socialist states" have generally established financial systems under which capitalists are simply replaced by government officials as the ones squeezing workers dry of surplus value. This is why the Marxist scholar Teinosuke Ōtani calls the Soviet Union an example of "state capitalism" instead.[23]

Many industries were nationalized in the Soviet Union, and in the realm of agriculture, there were both *kolkhozes*—collectively run farms—and *sovkhozes*—state-run farms; all of these were managed as parts of the planned economy. Yet these industries and agricultural production sites still bought and sold commodities for money in the marketplace.

The main goal of all these industries was still the maximization of surplus value—in other words, the valorization of capital. We can see how the essential characteristics of capitalism existed even in the Soviet Union.

Additionally, workers were barred from managing the mode of production themselves. This means that they had to sell their labor-power to a nationalized enterprise or state-run farm as wage workers producing surplus value for these government entities. Workers were as entrapped by the two senses of "freedom" discussed in chapter 2 as ever—their situation was essentially identical to what it would have been under capitalism, save for the "nationalization" at the apex.

While employment was comparatively stable in the Soviet Union, workers were still unable to change jobs or move. They were also restricted in terms of what commodities they could buy; however this isn't necessarily a consequence of socialism. Rather, it was a circumscribed freedom borne of bloated state control. In such a circumstance, the strict enforcement of price controls and the planned economy made innovation difficult, and workers lost motivation. It was in effect an inefficient version of capitalism that could not compete with the US.

There were attempts to redress this asymmetry with the West, including the Kosygin Reforms of 1965 and Perestroika in the late 1980s, that brought the economy even closer to the general form of capitalism. But these half measures weren't nearly enough to achieve Soviet goals, and in 1991, the Soviet Union ended up collapsing.

In China, on the other hand, the economic reforms begun under the leadership of Deng Xiaoping could be considered a success. But even so, they created a society that is by no means an improvement on capitalism.

"POLITICAL CAPITALISM" IN SOCIALISM'S CLOTHING

We can see similar problems in states liberated from colonization or occupation. Many such states, such as Vietnam, Cuba, and Indonesia, aimed to industrialize under the banner of socialism after gaining independence but ultimately evolved into antidemocratic "developing" dictatorships.

In these states, we witnessed the same system of officials using the fruits of "surplus labor" exceeding the immediate needs of the populace to reinvest in further industrialization. This is done under the theory that workers themselves "want" their surplus labor to be used this way for the development of the nation, and thus it cannot be a form of exploitation. But the fact of the matter is that these nationalized planned economies are guided by only a handful of top government officials, which means these systems must be understood as having been founded on exploitation.

Immanuel Wallerstein, the premier Marxist sociologist of the twentieth century, observed that once capitalism rose to the status of a "world system," what states like China, the Soviet Union, and various African nations attempted—modernization and the promotion of economic growth—turned out to be nothing more than national development by alternative means.[24]

These countries also have commodities, money economies, and capital; they also have exploited workers. In this sense, the countries claiming to be socialist in the twentieth century have been, in reality, simply dictatorships whose "socialism" fails to benefit workers. They're more properly thought of as practicing a form of "state capitalism," one that puts the economy under the control of government officials rather than capitalists.

The size of the government and the proportion of private to nationalized ownership are irrelevant to whether or not a society is capitalist. What matters is whether or not surplus value is being exploited with the aim of valorizing capital. This is why neoliberalism—a system of thought that argues on behalf of shrinking government and allowing society to be determined by the logic of the marketplace—represents only one side of the issue. China, whose government actively intervenes in the marketplace to make sure it functions smoothly, could never be said to be neoliberal, but it is nonetheless the world's preeminent *capitalist* state.

The economist Branko Milanovic recently called the US and Europe examples of "liberal meritocratic capitalism," as opposed to China, which he called an example of "political capitalism."[25] But the nature of "political capitalism" is really no different than what we've been calling "state capitalism." The fact is, both the Soviet Union and China have long been examples of "political capitalism."

NATIONALIZED OWNERSHIP ISN'T NECESSARILY COMMUNAL OWNERSHIP

Thus far we've detailed the common belief that the essential characteristic of capitalism is the private ownership of both the mode of production and the fruits of that production, which implies we could transition to socialism just by abolishing private property. Yet even when nationalization progresses, workers still end up exploited by cruel conditions imposed by the need to increase the value of capital, and high quantities of commodities are still traded in the marketplace within a money economy. Further, the domination of government officials leads to the rejection of democracy and the reckless expansion of state power.

If socialism really were nothing more than a synonym for nationalization, then of course no one would be attracted to it. But in fact, nationalization isn't socialism—it's *state capitalism*. In other words, when we talk about socialism, we shouldn't feel compelled to defend places like China, the Soviet Union, or "developing" dictatorships that pursue policies of nationalization and modernization at the expense of workers' rights and freedoms.

It feels intuitive to understand the private ownership of the mode of production as the foundational problem with capitalism, but this is a superficial understanding of how capitalism works. Let's recall the weight Marx put on "labor" in his discussion.

It's true that everyday life in a capitalist society involves the endless exchange of money and goods within a market economy founded on a system of private property ownership. But capitalism's essence resides in what lies beneath the fair exchange of commodities in the marketplace: the production of surplus value through the exploitation of workers.

It's not enough to try to solve this problem simply by nationalizing private industry. If we conceive of exploitation as merely a problem of *who owns* the mode of production rather than as a problem of *labor*, we risk making the same mistakes as the Soviet Union. What's truly necessary to overcome capitalism is the development of a *form of labor* free of exploitation.

As we've seen, wage labor premised on the exploitation of the surplus labor of others exists within industry even when it's nationalized. Nationalized ownership is not the realization of communal ownership. Rather, it produces a ruling class of government officials like the *nomenklatura* of the Soviet Union where both the mode of production and the system of ownership of the fruits of that production are entirely controlled by others, completely disconnecting the workers from their will.

FREE EDUCATION AND HEALTH CARE IN GERMANY

So why is it that the discourse tying nationalization to socialism remains so strongly rooted? One reason is that the transition to nationalization is accomplished through political power. Indeed, a defining feature of state capitalism is that it solves economic problems through political force and the power of the state rather than through the self-determination of the workers.

Marx was aware of this danger. He warned that embracing a superficial understanding of capitalism can give rise to the "juridical illusion" that all that needs to be done to change the law is to take the reins of power via revolution or election (766). When we look at extant examples of "socialism" we can see that they've succumbed to precisely this illusion.

According to Marx, more fundamental than any law or system is the power wielded by commodities and money to dominate humanity. Recall that Marx called this inversion of the relation between people and things "reification" (see chapter 1).

What I want to emphasize here is that Marx didn't see taking back the power of the state and instituting revolutionary change to the political system as the most important step in resisting capitalism; rather, he thought it was most essential to lessen the power of reification in our economic systems. This might seem like a tall order, but Marx argues that the more we live independently of money and commodities, the more opportunities we'd have to make choices for ourselves in our day-to-day lives.

For example, higher education in Japan is very expensive. Classes at a private university can cost around 930,000 yen [$5,913.33 USD] a year. With matriculation and other fees added in, that can come to 4,000,000 yen [$25,412.16 USD] over four years. Even state universities are becoming more expensive, costing around 540,000 yen [$3,430.64 USD] a year. And unless you continue to live with your parents, there's also the cost of rent and other daily expenses. Students thus end up taking out loans as "scholarships" that must be paid back with interest, racking up debts by graduation. Students often feel pressure to graduate within four years and immediately get a job to work off the debt, working as hard as they can at whatever job they get, unable to even entertain the idea of quitting.

The situation is quite different in Germany, where I was an exchange student. There were a few who graduated in four years—it was typical to take six. There were doctoral students who'd been in school upward of twenty years, which shocked me when I first arrived.

How could this be? Because tuition is free. For about 120 euros [$140 USD] a semester, you can get a student ID card that entitles you to unlimited use of public transportation like buses and trains. You can also use it to eat cheaply at the school dining halls and receive discounts at museums and concerts. You can go see the Berlin Philharmonic for about fifteen euros! Universities also offer affordable student housing.

It's not just tuition. Health care is also, in principle, free in Germany, with warm, attentive care services. There's support and job training for the unemployed, and daycare is also free. There's no need to amass 20 million yen just to be able to retire. Taken together, this safety net allows citizens to feel less pressure to stay in a job they're dissatisfied by.

This is precisely what welfare state scholar Gøsta Esping-Andersen calls "decommodification."[26] The more the goods (housing, public parks) and services (education, health care, public transport) necessary for daily life are available free of charge, the more a society is "decommodified." Rather than having to use money to procure these necessities in the marketplace, people can access them directly as simply what they are—health care, education, etc.—rather than as commodities.

By providing citizens with benefits in kind, a state can lessen their burden to continue working solely for an income. A welfare state is obviously also still a capitalist state. But through decommodification, the power of reification can be curtailed considerably.

ASSOCIATION PREDATES NATIONALIZATION

Education, health care, and the like were free in the Soviet Union, too, so it's often hard to recognize the difference between it and a welfare state. But one key difference was the Soviet prioritization of nationalization above all else. Welfare states, on the other hand, emerged out of social movements aiming to weaken the power of reification via what Marx calls "association."

Marx himself only rarely uses terms like "communism" and "socialism." The term he uses most frequently when discussing a postcapitalist society is "association."

Labor unions, workers' co-ops, labor parties—these are all forms of association. Today, we'd include entities like NGOs and NPOs. Marx was not calling for a government official–dominated society like the Soviet Union, but rather for a democratic society founded on freely given mutual aid and fellowship.

We can see the importance of association by looking at the historical process by which the various services offered by the welfare state were established. For example, unemployment insurance arose when workers took it upon themselves to pool a portion of their wages to support fellow workers when they lost their jobs. After all, solidarity within a labor movement would be undermined if there were too many people so desperate for work they'd take any job at all, no matter how low paying.

From social welfare and pensions to public libraries and health care, when we trace common forms of social support back to their origins, we find labor unions, neighborhood associations, cooperatives, and so on. Faced with the power of capital, people took it upon themselves to create mutual aid systems to protect and enrich their lives.

In other words, these forms of social support all arose from *associations* that aimed to decommodify society; starting in the beginning of the twentieth century, this support was taken up by the state, which offered it to all of its citizens using funds that no longer came directly from workers, but rather from taxes. This is a reversal of the process by which state capitalism emerges, in which nationalization comes first, driven by political parties and government officials. By contrast, in welfare states, universal services offered by the state emerge only *after* their creation and development by associations.

Because associations form the basis of the economy, social support doesn't need to depend on the supervision or administration of the state to exist. In Germany, there was no nationally mandated minimum wage until very recently. German labor unions are organized by industry, not by company as they are in Japan and elsewhere. (IG Metall is the metalworkers' union, Ver.di is the service workers' union, and so on.)

Labor-management agreements put in place across every enterprise within an industry were sufficient to keep minimum wages in place, obviating the need for the state to intervene in the marketplace to do so.

However you look at it, welfare states contain elements that overlap with Marx's vision for a future society, especially relating to decommodification and limiting the power of reification. They're closer to this vision in terms of association as well, certainly more so than "socialist" states like China and formerly the Soviet Union, which forbid labor unions while government officials monopolize decision-making in the name of nationalization.

THE "JURIDICAL ILLUSION" OF BASIC INCOME

Marx's criticisms of state capitalism and "juridical illusion" are more important today than ever. These phenomena are hardly things of the past. As labor movements stagnate and association weakens, reforms are being proposed once more to use state power to ameliorate the effects of capitalism.

Basic income (BI), which has gained widespread popularity since the beginning of the 2000s, is symbolic of this kind of juridical illusion. At first blush, it seems quite bold to pass a law that gives every citizen money. If we all automatically received enough to live on, we would no longer have to work at jobs we hate. We could have more satisfying careers, enjoy more free time, lead richer lives—at least this is the suggestion.

It may seem like BI is a miracle cure, able to bring the patient back from the brink of death. Yet distributing two or three hundred dollars a month per citizen is essentially pointless if it means cutting pensions and social welfare spending. On the other hand, giving something more like a thousand dollars a month would put a considerable financial burden on big business and the wealthy elite.

Capital would therefore do anything it could to prevent such a tax hike from happening. Multinational corporations would surely threaten to move their business somewhere with less of a tax burden if, say, the Japanese government were to announce that it was going to raise taxes to fund a BI program. And if they made good on these threats, tax revenue would go down and stock prices would crash. This is what is known as a *capital strike*.

Capital has the freedom to bypass national boundaries and invest in what it likes, where it likes; this is the origin of the superior power it wields over states and workers, who lack such freedom of movement. This freedom provides the security that would allow a capital strike to take place. A nation that wishes to institute BI must be able to withstand such a strike, and will need social movements of commensurate power to provide their own type of security for the state and its workers.

But if social movements really had this much power, there would be other ways to effect social change besides getting the state to hand out money. For example, they could make social services like health care, higher education, homecare, daycare, and public transportation free—in other words, they could decommodify society.

BI's popularity has taken place against a background of weakened labor movements, widespread precarity, and increasing numbers of low-paying jobs. BI attempts to address the failure of labor movements to guarantee workers' livelihoods by using the power of the state to distribute money instead.

It's indeed true that if workers received money every month, they would have less precarity in their lives; BI could even fortify the position of the working class. But such a policy does nothing to address the mode of production, which means that it does little to lessen the power wielded by capital. It's unclear how those calling for BI would have enough power to stand up to a capital strike.

As a policy proposal, BI rather naïvely takes for granted the power money wields in the present moment. Doesn't this mean that we'd continue to be as manipulated by money and commodities as we are now? It's a policy that does nothing to lessen the power of reification.

Marx, by contrast, sought to reduce reification by calling for the transition to a society wherein money and commodities wield no power at all. Naturally, we can never use money to achieve such a goal. The only way to escape the power of money is by expanding our ability to live without using money. The only way to do this is through the power of association.

THOMAS PIKETTY, MMT, AND THEIR SHARED BLIND SPOT

A similar kind of "juridical illusion" can be found in the reforms to the tax system proposed by the French economist Thomas Piketty, bestselling author of *Capital in the Twenty-First Century.*[27] Recently, Piketty has called for "socialism," through the large-scale redistribution of resources via dramatic raises in income taxes, corporate taxes, and inheritance taxes. He proposes raising maximum income and inheritance tax rates to 90 percent, then using this money as principal to give every adult upward of one hundred thousand dollars each. This kind of redistribution would certainly stabilize and enrich the lives of the average citizen.[28]

But it's just as obvious that capital would despise such a proposal and do everything in its power to thwart it. Just as in the case of BI, it's unclear from Piketty's proposal where the power to effect such a large-scale change would come from in the face of a capital strike.

In the end, top-down proposals like Piketty's that depend on good-hearted elites to redesign systems to benefit society are juridical illusions doomed to failure. We must find a way to foster the power of association so we can effectively stand up to a capital strike, and Piketty's proposal lacks a clear vision for how to create a movement that can support such a sweeping reform.

Modern Monetary Theory (MMT), a school of anti-austerity thought that has been slowly gaining popularity, suffers from the same problem. MMT is distinguished by its contention that since any state that issues its own currency can run up a budget deficit without defaulting, state spending should never be cut even if it makes the deficit go up, as long as inflation remains under control.

The large-scale stimulus spending advocated by MMT comes with a "Job Guarantee" program to safeguard the lifestyles of workers: The government will create jobs that pay an adequate minimum wage that allows all citizens to work in the field of their choice. The idea is that economic growth will be fostered by the active creation of jobs necessary for the transition to a more environmentally friendly, sustainable society.

Will such a scheme really work? Even if this represents an expansionary fiscal policy, the specific gravity of the economy will shift to public investment, which means that it will ultimately be the government who decides what to invest in. Capital is guaranteed to object to this. The foundation of capital's power is its ability to decide where it should be invested—it will go to any length to resist measures that might undermine that.

But controlling capital through public investment is key to making MMT work. If it were simply a matter of spreading money around, funds may well end up being used to expand the military or on wasteful public works projects rather than to create employment opportunities that both benefit society and protect the environment. Further, the process of handing out money might generate opportunities to charge and earn interest, which would end up shifting profit making back to big business once again.

On the other hand, as the government's intervention into the marketplace grows and regulations aiming to reduce carbon emissions and guarantee human rights strengthen, capital's resistance will grow stronger as well. Capital may begin to withdraw from domestic investment, there might be a currency sell-off, and the pressure of inflation may go up. In this scenario it becomes necessary to introduce austerity measures by raising taxes and interest rates. Ultimately, MMT economic policies lack the power to combat the risk of a capital strike.

A state needs sufficient power of association to withstand a capital strike even when trying to effect large-scale top-down policy changes. And association requires that workers hold the power to determine the terms of production: where investment will go, how work will be conducted, and so on.

Revolutionizing the site of production like this is obviously a tall order. But that's no reason to shy away from the fundamental task before us: to change the balance of power between capital and wage labor. MMT, Piketty, and BI all lack a strong perspective on how to create associations powerful enough to generate and support this kind of change.

This is no accident. Tax reform, BI, and MMT are all top-down proposals, conceptions of political change symptomatic of an age without class struggle. They're captured by the juridical illusion that prioritizes policy and law above all else.

By contrast, Marx's unique perspective, which ties together reification, association, and class struggle, allows us to greatly expand our theory and practice so we can bring about truly bold societal change.

TOWARD A BOTTOM-UP TRANSFORMATION OF SOCIETY

Marx rejected the idea that society could be changed for the better through solely top-down policy changes. (In this way, his thought differs greatly from that of Thomas More, who proposed a planned Utopia.)

This point is extremely important. Placing decommodification through association at the center of the strategy for social change allows it to differ in essential ways from the vision proposed by the revolutions of the twentieth century, of which the Russian Revolution looms large in the public imagination. This is a major shift from a top-down to a bottom-up model.

We can see this shift in Marx's own views on revolution. Marx was still quite young when he cowrote the *Communist Manifesto* in 1848, where he advocates for using the financial crisis as an opportunity to seize the power of the state—to nationalize the means of production and install a "dictatorship of the proletariat." But in *Capital*, this discussion plays out very differently; we no longer see the same wish for a crisis to precipitate change (though it's also true that he hasn't let go of the idea of the dictatorship of the proletariat).

Rather, as we've discussed in chapters 2 and 3 of this book, the Marx of *Capital* places great importance on shortening work hours and allowing workers to retain their skills. It's a book of revolution, yet it focuses much of its attention on reforms set in motion through association *within* capitalism.

Beneath this shift is Marx's personal experience with the difficulties of revolution. The Marx of the *Communist Manifesto* was optimistic, imagining that the immiseration of the proletariat and the coming financial crisis would spark a revolution that would establish a new, socialist structure for society. But the revolution triggered by the workers' revolt in 1848 failed considerably, and capitalism caught a second wind and recovered. The financial crisis of 1857 played out in much the same way.

Faced with capitalism's tenacity, Marx rigorously investigated the source of its power. This led him to research economics, the result of which was *Capital*; in it, Marx sheds his former optimism about revolution, instead arguing on behalf of revising capitalism with an eye toward revolutionizing it.

Marx thus began to prioritize shortened work hours over increased wages, an essential differentiation from the point of view of reification. Raising wages is of course also important, but it ultimately fails to liberate workers from their own desire to work longer hours and accumulate more money. In fact, it would make workers even more dependent on money—after all, desire is by nature limitless.

The welfare states of western Europe have indeed made work hours shorter. In France, for example, the workweek is thirty-five hours compared to the typical forty (or more) hours a week that Americans and Japanese workers endure. This reduction offers more free time to do things other than work. However, if stores remain open even on Sundays, workers are still being subsumed by capitalism in their time off. This is why in certain welfare states there are also regulations stipulating that with the exception of restaurants and museums, businesses like department stores, shopping malls, and supermarkets are all to be closed on Sunday.

When stores are closed, it becomes nearly impossible to engage in capitalistic consumption. The term "window shopping" is not, as some believe, a term for looking through shop windows at luxury goods you can't afford; rather, it's what you do when you have no choice but to content yourself with just looking when the shop is closed on Sundays.

When we cannot shop we find other ways to spend our free time. We might read a book in a café, or engage in political discussions. We might join a local sports team and spend the day playing soccer; we might grow a garden or an orchard. We might volunteer, or join a protest.

Free time borne of decommodification creates the space for skills and activities unrelated to capitalism to bloom. They escape capture by cost-performance thinking, fostering the richness of social prosperity.

THE LIMITATIONS OF THE WELFARE STATE

That said, welfare states haven't completely solved the problems of capitalism. Even if we concede that partial reforms can and do occur within a capitalist framework, we can still consider the system overall and see that it is, indeed, still capitalism. Various contradictions arise. I'd like to close this chapter by pointing out four limitations of the welfare state.

The first of these is the expansion of bureaucracy. The goods and services offered by the welfare state may have begun as outgrowths of association, but as they're taken over and administered by the state, the bureaucracy overseeing this administration expands greatly. Though the process is different than what occurred in the Soviet Union, inefficiency and overreach by government officials nonetheless still arise.

We can see a related shift in the people's point of view—from voluntary participation in associations to more passively receiving benefits offered by the state and leaving political organizing up to the leaders of big labor unions. The working class must then find common ground with the capitalist class; the result is that workers internalize capitalist values and perspectives as their own, all while undermining the power of association.

The redistributive policies of the welfare state were originally made possible by high-speed economic growth. The extraction of cheap foodstuffs and other resources from the Global South benefited not just capitalists, but workers as well.

High-speed economic growth enabled the average household to buy things like cars, air conditioners, and television sets, and in that sense, enriched the average person's life. As a result, workers' desires became aligned with those of capitalists, uniting them in the quest to improve their lifestyles by extracting resources from the disadvantaged of the world.

This is where the second limitation of the welfare state becomes clear: the problem of inequality between the Global North and South. While workers' lives may appear to be less commodified, true decommodification can only occur within the countries of the Global North (or within supra-national entities like the EU) and can ultimately only be sustained by the large-scale extraction of resources from former colonies and other parts of the Global South. Yet workers don't criticize this exploitation of those in weaker and more precarious positions than themselves—rather, they are satisfied to receive the gifts bestowed to them by capitalism. This is the phenomenon German sociologists Ulrich Brand and Markus Wissen have dubbed the "Imperial Mode of Living."[29]

This is a mode of living based on exploitation and externalization, which brings us to the third limitation of the welfare state. It's not just humans that are exploited to support the Imperial Mode of Living—nature is as well. As material improvements to working-class lifestyles took precedence over other considerations, these lifestyles became defined by mass production and consumption at the expense of the natural world. As problems like environmental pollution and landscape destruction became impossible to ignore, the welfare state became the object of justified criticism by the environmentalist movement.

The fourth limitation of the welfare state is its patriarchal character—that is, its reproduction of gender inequality. The labor movement has ultimately devolved into a movement by and for men. The care work and reproductive labor necessary to allow men to go to work every day—childcare, cooking, washing, and so on—enables capitalism to function at all, yet it remains relegated to the realm of unpaid work. The welfare state does shoulder some of the burden of reproductive labor by providing daycare and other forms of care work and in this way works to "degender" this labor, but by and large such work is still foisted on women.

All of these limitations to the welfare state must be faced head-on. Even as we expose the contradictions of neoliberalism, we must also resist returning uncritically to the welfare-state model.

The high-speed economic growth that enabled the initial formation of the welfare state is no longer attainable in the Global North. Planetary limits on resource extraction like climate change restrict how much more exploitation and externalization can be done at all. With these critiques in mind, a movement that's patriarchal, bureaucratic, and centered on the Global North lacks the breadth necessary to truly change society.

In other words, it's not just class that's at issue—it's also gender, the environment, race, and so on. We must think of a new way to decommodify, a new way to associate. I believe this can be done through renewing the commons through the "degrowth communism" conceptualized by Marx near the end of his life.

6

WHO SAID COMMUNISM'S IMPOSSIBLE?

WHY DIDN'T MARX CONCRETELY DESCRIBE THE SOCIETY TO COME?

While Marx imagined that there could be a society after capitalism, the exact shape of that society was quite vague. Imagining such a thing in concrete terms is an incredibly difficult thing to do in the first place; furthermore, *Capital* was still unfinished when Marx passed away.

But there's another reason his vision lacked specificity—Marx knew that there was a risk that when we try to imagine this future society, we'll end up uncritically reproducing our current value systems and "common sense." We might make the mistake of envisioning supposedly revolutionary forms of work, freedom, and equality that ultimately are still based on the desires and assumptions (about race and gender, for example) that characterize the capitalist world we already live in. Marx thought the future society should be created by those actually living in the future, and therefore he refrained from being too specific and concrete in his own rendering of it. We cannot depend on Marx to imagine it for us—we must try, through trial and error, to conjure it ourselves.

But many Marxist scholars, fooled by the presence of Soviet-style "socialism," failed to think seriously about what shape the postcapitalist world could take. After the Soviet Union's collapse, it became even harder to imagine any sort of society besides a capitalistic one. Today, it feels nearly impossible to articulate any sort of system that might replace capitalism.

As if to evade this question, starting in the 2000s it became fashionable for scholars to read Marx's *Capital* simply as a critique of capitalism, allowing the book to be severed from debates about communism itself and instead serve as a way to analyze the problems caused by capitalism: death by overwork, financial crises, environmental destruction, and so on. But aren't we missing something by reading it this way?

Marx devoted himself to writing *Capital* at the expense of his family and his own health. It was the culmination of a lifelong project to create a new society that had overcome capitalism. But so many people now call themselves Marxists without actually calling for communism. Instead, they content themselves with criticizing capitalism's "excesses"—a form of Marxism that, to me, stops at a mere critique of neoliberalism. In this way, Marxism loses its uniqueness and thus its *raison d'être*, which will lead to its inevitable decline as a form of thought. It is essential that we read *Capital* specifically to imagine the utopia called communism.

WHAT ENDED UP LEFT OUT OF *CAPITAL*

As I touched on in the previous chapter, Marx's thought changed radically over time. In my discussion here, I focus on Marx's late-period thought, just before the end of his life.

Marx was writing *Capital* in London, where he'd exiled himself; even after the publication of the first volume in 1867, he'd visit the British Museum's reading room every day to continue his research. However, he ultimately published very little due to his worsening health.

To home in on the contours of Marx's later thought, we must not only read the published version of *Capital* edited by Engels, but also the manuscript versions of it written by Marx himself, as well as his preparatory research notes, as it's necessary to look at what he was researching as he was preparing what was to become *Capital*'s second and third volumes. And this endeavor has, in recent years, finally become possible.

As discussed earlier, the new collection of the complete works of Marx and Engels—the project known as MEGA—has allowed researchers to access the excerpts Marx copied into his notes, giving readers new clues to what he was thinking about in his final years. There are, for example, extensive passages written by scholars like Justus von Liebig and Karl Fraas, passages that show the depth of Marx's concern about environmental issues like deforestation and resource exhaustion.

But this isn't the only aspect of Marx's late-period thought that's been brought to light by the MEGA project. Marx's notes chronicling his research into metabolism illuminate a major reversal in his thought during this period.

"EQUALITY" IN THE ANCIENT COMMUNE

Along with ecology, Marx was ardently pursuing another subject during this period: the ancient communal societies that existed in precapitalist western Europe and elsewhere. He first read about the *Markgenossenschaft* (mark-associations) of the ancient Germanic peoples, but his interest widened to include communal societies in ancient Rome, the indigenous Iroquois people of North America, the *mir* of Russia, and forms of communal society in South America and India.

Initially, his interest was sparked by the discovery that in an "archaic" commune like the *Markgenossenschaft*, land was treated as a communal resource where people could live in a state of "equality."

Marx took detailed notes on German legal historian Georg Ludwig von Maurer's book about these Germanic communal societies; he returned again and again to the subject over the course of multiple years. As he was assembling his notes on the egalitarian aspects of these communes, he reported finding—just as he had with Fraas—"socialist tendencies" in Maurer's study of the early Germanic societies. In both men's work Marx saw historical evidence that communities didn't have to be stratified by social systems like capitalism to function.

As he conducted his research into communal societies and the natural sciences at the same time, Marx became increasingly aware of the strong connection between *sustainability* in the natural world and *equality* in the human one.

The uneven distribution of wealth creates political power—relations of domination and submission—and the codification and strengthening of this unevenness leads to the exploitation of both humans and nature. When these resources inevitably become exhausted, a battle to steal what remains from one another commences. There's no way for society to thrive and prosper under such conditions.

Precapitalist communal societies avoided this sort of unevenness in various ways: through traditions and religious customs, through the communal ownership of land, through the distribution of resources via lottery, and so on.

In a communal society, you can't impose your views unilaterally on others. People who don't like you might decide to simply leave the commune; conversely, if you find that you're disliked by everyone else, you might find you have to leave the commune yourself. This is the sense in which a society without domination becomes "equal." Contrary to the stereotypical imagination of communes, individuals in such a society are compelled to act and make decisions on their own behalf rather than submit to the will of others in a master/servant dynamic.

TURNING AWAY FROM HISTORICAL MATERIALISM

But how were these archaic communal societies able to foster both sustainability and equality? They were able to govern the metabolic relation between humans and nature in an entirely different way than capitalism demands. The ultimate goal of Marx's diligent research into these early societies was to illuminate this argument.

As we saw in chapter 3, Keynes observed that the lifestyles of ordinary people didn't change dramatically until the beginning of the eighteenth century; by the same token, production took place in communal societies in the same way year after year. The same sorts of crops were planted the same way at the same time. Sowing and harvesting were tied to festivals and rituals and performed by everyone. To use the categories defined by the economist Robert Heilbroner, communal societies operated according to a Traditional economic system.[30]

To prevent wealth from being monopolized by the few and thus becoming the object of competition and theft, the scale of production and the possession of personal individual wealth were both strictly regulated, meaning these societies could realize a "steady-state economy" that functioned while keeping levels of production, consumption, population, and capital all unchanged.

Because trade with outsiders was limited, these societies could develop a nearly self-sufficient circular economy. While there were no dramatic rises in productivity, there weren't any undue burdens such as soil exhaustion imposed on nature in order to sustain the functioning of human society.

But these types of communal societies weren't all relegated to the distant past. Marx discovered some that still existed during his own time: Russian agricultural communes known as the *mir*.

Marx's collaboration with Russian revolutionaries during the preparation of the Russian translation of *Capital* inspired his interest in these communes. Marx taught himself Russian during the process, which enabled him to read some of the research materials in their original language.

Marx also discovered that the Russian *mir* had a certain "economic superiority" compared to western European societies, even though they were unmodernized agricultural communes. The *mir* achieved stability through communal forms of labor and ownership from which arose sustainability and equality. This discovery represented an extremely important shift in Marx's thought.

According to the conventional understanding of Marx, productivity is the fundamental force driving history to ever-higher levels of development. This is a viewpoint known as "historical materialism."

But this theory necessarily posits that the most technologically developed countries are also the most advanced regions in the world. By this logic, "primitive" peoples must be "civilized" through the introduction of capitalism, thereby justifying colonialism. Historical materialism has thus often been interpreted as a wellspring of racism. And indeed, Marx has routinely been critiqued as a Eurocentric thinker since the turn of the twentieth century.

Yet Marx's own research into non-Western communal societies in Russia and elsewhere led him to abandon this historical perspective. Instead, he held these communal societies in high esteem, asserting that they'd preserved forms of equality and sustainability that had been lost in western Europe. He even went so far as to say that they could act as a foundation for communism.

TOWARD DEGROWTH COMMUNISM

The new historical perspective Marx developed during his later years is articulated most effectively in the drafts of a letter he wrote in 1881 to the Russian revolutionary Vera Zasulich. He writes:

> Russia is the sole European country where the "agricultural commune" has kept going on a nationwide scale up to the present day . . . On the one hand, the common ownership of land allows it to transform individualist farming in parcels directly and gradually into collective farming, and the Russian peasants are already practicing it in the undivided grasslands; the physical lie of the land invites mechanical cultivation

> on a large scale; the peasant's familiarity with the contract of *artel* facilitates the transition from parcel labor to cooperative labor; and, finally, Russian society, which has so long lived at his expense, owes him the necessary advances for such a transition. On the other hand, the contemporaneity of western production, which dominates the world market, allows Russia to incorporate in the commune all the positive acquisitions devised by the capitalist system without passing through its Caudine Forks . . . it can become a direct point of departure for the economic system towards which modern society tends; it can turn over a new leaf without beginning by committing suicide.[31]

Here, Marx is saying that Russia can avoid going down the degrading path of capitalism, which he compares to the "Caudine Forks"—the name for a valley in which Roman soldiers, during the Samnite War, found themselves trapped and then were forced to humiliate themselves in front of the enemy by passing beneath a symbolic wooden yoke while stripped to the waist. Using the communal nature of the *mir* as a starting point, Russia could skip this struggle and leap directly into communism.

In other words, there was no need to destroy the commune through the introduction of capitalism. If the Russian commune was able to skillfully incorporate the fruits of western

European capitalism without the forced imposition of this sort of system from the outside, then they could, by their own power, develop their own version of communism—the "economic system towards which modern society tends." This argument shows how Marx had in part shed his early Eurocentric and progressive view of history, signaling an enormous shift in his thought.

Building on this, I'd like to draw our attention to another passage from the same letter:

> Today, [capitalist production] faces a social system which, both in Western Europe and the United States, is in conflict with science, with the popular masses, and with the very productive forces that it generates. In a word, it finds it in a crisis which will only end in its elimination, in the return of modern societies to the "archaic" type of communal property. In the words of an American writer who, supported in his work by the Washington government, is not at all to be suspected of revolutionary tendencies, "the new system" to which modern society is tending "will be a revival, in a superior form, of an archaic social type."[32]

Here, Marx asserts that in order to solve the present crisis, western Europe must "return" to the type of communal society exemplified by the Russian *mir*. But this is not a call to reject advancements in technology and civilization and go back to nature—Marx intentionally includes the quote indicating that this new system will instead be a revival of "a superior form" of communal society.

These excerpts make clear that Marx no longer believed that uncontrolled modernization should be pursued simply by adopting capitalism. After all, it's the technological advancement driven by competition and the endless quest to valorize capital that opened up the "irreparable rift" in the metabolic relation of humans and nature in the first place.

Recall that communal societies were able to maintain a stable, sustainable social arrangement through the continued observance of traditions. Marx, in the later period of his life, believed that western Europe must transition to this sort of arrangement.

The form of richness Marx sought during his research into the Russian *mir* can't be measured by looking at the GDP or the sum total of individually held assets. How can the GDP measure happiness, or the cleanliness of drinking water? Rather, Marx demanded that we return richness to societal wealth by excising our zeal for modernization and economic growth and replacing it with a renewed emphasis on the interdependence of humankind and the natural world.

Unlimited economic growth is not necessary to the realization of this "superior form" of communal society. Neither is an endless rise in productivity levels. That's why I refer to the form of communism Marx conceived of near the end of his life as "*degrowth* communism." This—not the ambition to become a global superpower found in the Soviet Union or China—is what gives way to the possibility of fashioning a truly postcapitalist society.

"FROM EACH ACCORDING TO HIS ABILITIES, TO EACH ACCORDING TO HIS NEEDS!"

Unfortunately, though, Marx passed away before he could articulate a fully developed vision of degrowth communism. It is up to those of us living in the twenty-first century to think in tandem with, while simultaneously moving beyond, Marx's thinking in order to create our own vision of the new society to come. (For more detail on this point, please look at my previous book, *Slow Down*.)

But for now, let's look at what Marx actually did say about the society of the future. A passage from near the end of *Capital*'s first volume gives us a useful starting point:

> This is the negation of the negation. It does not re-establish private property, but it does indeed establish individual property on the basis of the achievements of the capitalist era: namely co-operation and the possession in common of the land and the means of production produced by labor itself. (929)

The word translated as "land" is *Erde*. But here Marx uses the word in the broadest possible way, referring to the entirety of the natural world. When we hear the word "land," the first image that comes to mind is likely a piece of real estate, but Marx is conceptualizing not just the land itself but also the vegetation growing in its soil, the streams and rivers flowing through it, and the resources hidden beneath its surface.

The "negation of the negation" Marx refers to is the negation of capital's negation, through primitive accumulation, of nature and of workers' control of the means of production, resulting in the exploitation and monopolization of both by capital. Marx calls for liberation through the restoration of nature and the means of production into things held "in common"—that is, as community property. No one will own these things as an individual—that is, as *private property*—but rather everyone will, communally. The land—the planet—belongs to no one. It's not something for industry, or for those of us living in the Global North, to exploit and destroy at will.

Marx, in the manuscripts that became *Capital*'s third volume, explores this question directly. (The translation here uses the word "earth," not "land.")

> From the standpoint of a higher socio-economic formation, the private property of particular individuals in the earth will appear just as absurd as the private property of one man in other men. Even an entire society, a nation, or all simultaneously existing societies taken together, are not the owners of the earth. They are simply its possessors, its beneficiaries,

> and have to bequeath it in an improved state to succeeding generations, as *boni patres familias.* (911)

The water, the forestland, the riches beneath the surface of the earth—all these basic forms of wealth must be managed sustainably, not by the state or the marketplace, but by everyone, through association, as a form of commons.

Let's use the cultivation of apples as an example. Once harvested and distributed to everyone, each apple becomes each person's individual property. But the orchard and the tools necessary to cultivate it—as well as the knowledge guiding that cultivation—are all types of communal property; they belong to everyone. There's no need to convert them into private property. All the forms of wealth that capitalism has expropriated via enclosure and monopolization must be returned to everyone as producers, and thus be communally owned and managed at that level.

The future society envisioned by Marx is none other than the *restoration of the commons.* A society founded on shared access to the commons is the true endpoint of "communism." In his last years, Marx imagined a world in which societal wealth was no longer defined by commodities—rather, the economy functioned in a steady state based on equality and sustainability, with everyone participating in its autonomous self-governance.

How, exactly, would this wealth be shared and managed? "From each according to his abilities, to each according to his needs!"—this is the famous line from *The Critique of the Gotha*

Program, a treatise written by Marx near the end of his life, in 1875.[33] To spell out the slogan's implication, Marx called for everyone in society to use their respective skills to contribute whatever they can to the collective's prosperity; through this process, every individual will receive what they need to survive and thrive.

This does not and cannot happen under capitalism, which is dominated by commodity exchange, or the exchange of goods based on equivalence in value. There's no lending or borrowing between parties, just a simple exchange and a fare-thee-well. It's a cold and disconnected system.

Meanwhile in a truly communist society, people support each other, lending their skills and time to benefit the community as a form of gift-giving without the expectation of equivalent exchange. But this gift-giving isn't one-way—you take what you need as well. The food, land, tools, and knowledge that form the richness of that which is necessary to life become things that are shared via concrete practices of giving and receiving.

Mutuality and reciprocity. The world has many types of people in it. There are those who are bad at math but enjoy using their bodies, those who find it hard to articulate their thoughts aloud but excel at drawing pictures. There are those who hate to exercise but love to study. Skills differ between individuals, but for this very reason, society can become a place where everyone's strengths can be used to aid one another. Differences in skills are both natural and desirable. And when we need help, we should be allowed to ask for it.

Socialism is often mischaracterized as a system in which the state obscures and elides individual differences in favor of imposing a uniform, one-size-fits-all version of "equality" on its citizens. But there are alternative ways to create truly equal societies.

THE EXPERIENCE OF THE PARIS COMMUNE

In addition to his research into Russia's early communal societies, there was another major reason that Marx's thought underwent such a thoroughgoing change near the end of this life—the Paris Commune.

During the Franco-Prussian war, and following the elections in February of 1871, the new French government, led by Adolphe Thiers, attempted to make peace with Prussia by allowing Prussian troops to occupy Paris. The citizens of Paris, outraged, rose up in an armed rebellion, driving Thiers and his forces to flee Paris for Versailles. The self-governing revolutionary body that arose in the wake of his flight was the Paris Commune.

Though it lasted only two months before being suppressed by government military forces sent back to Paris from Versailles, the Paris Commune left its mark in history as the world's first autonomous worker-led government.

Marx was, at first, skeptical of the Parisians' revolt. It struck him as a premature stab at revolution—a failure that gave way to a huge number of casualties. Yet, as the hard work of the anarchists began to pay off and a worker-led form of autonomous governance was established, Marx's opinion changed. During those two months before its military suppression, the Commune represented the emergence of a genuinely new form of society in the heart of a capitalist country.

The Paris Commune demonstrated that it was possible to avoid the most common charge leveled against socialism: that it will lead to a strict conformity borne of domination by bureaucracy that ends up sacrificing freedom and democracy to the cause. The Commune, by contrast, overflowed with diversity as men and women from all walks of life joined together to create a peaceful, democratic self-governing body. Even foreigners were allowed to join and participate in the Commune.

The military and police forces, embodiments of state violence, were disbanded. Since the bureaucratic system had been dismantled, workers themselves became the architects of government as council members. In the hands of the Paris Commune, the governing council transitioned from being a body that only created laws to one that also administered them. Naturally there were still various bureaucratic functionaries left over from the previous government, but their pay was lowered to match that of the average worker; their positions were also transformed into elected ones, and as representatives of the public will, they could be recalled at any time. Finally, the clergy were "returned to seclusion"—that is, removed from public life, including the schools and churches they ran—and church-held assets were seized.

Major changes occurred at the site of labor as well. Arduous night shifts and child labor were abolished. Education was made free, secular, and available to all children. The influence of the church and state were removed from the classroom, and workers became teachers. The wage gap between men and women was abolished. School became the site of vocational education, and available to all regardless of differences in gender, class, or race; the goal was to overcome the separation of conception from execution.

In the Commune special classes disappeared and associations and cooperatives of all sorts were established one after another. This was the purest expression of both workers' democracy and the restoration of the commons. In Paris, a city that was the beating heart of capitalist society, mutual aid and gift giving spread, displacing the exchange of money and commodities in the pursuit of capital increase. Using this dramatic shift in the economy as a basis, an entirely new form of democratic governance emerged—not in the form of a state, but a commune.

Observing and experiencing this event changed Marx's theorization of communism substantially. For example, in the 1872 preface to the German edition of the *Communist Manifesto*, Marx incorporates a quotation from *The Civil War in*

France, his book on the Franco-Prussian wars and the subsequent Paris Commune:

> In view of the practical experience gained . . . in the Paris Commune, where the proletariat for the first time held political power for two whole months, this program has in some details become antiquated. One thing especially was proved by the Commune, viz., that "the working class cannot simply lay hold of the ready-made State machinery, and wield it for its own purposes."[34]

Here, Marx emphasizes how his thought had shifted since he first cowrote the *Manifesto* in 1848. The original version of the *Manifesto* reflected his belief that socialism could be established simply by seizing the power of the state through revolution. The experience of the Paris Commune, however, led Marx to argue that a truly egalitarian, democratic society could not be established simply by seizing state power but instead required a different, more dynamic path.

THE COMMUNE, OLD YET NEW

The Paris Commune influenced the way Marx thought about the Russian *mir* and other communal formations he'd been researching. In many ways, the Russian "communal society" and the Paris "commune" in western Europe had deep resonances with each other.

From today's perspective, it's easy to find faults in the understanding of ancient societies and the Russian communes that Marx gleaned from the work of Maurer and Fraas. There are historical inaccuracies and Eurocentric ideas in Marx's views on non-Western societies. But this shouldn't lead us to conclude that Marx spent the last years of his life idealizing premodern foreign societies because he misunderstood them.

What Marx saw in these communal societies of the past wasn't pure fiction. In the *pre*-capitalist social formations of Russia he discovered systems and values similar to the "post-capitalism" that emerged in his own time in the very heart of modern capitalism. Though separated by time and geography, the two examples resonated with one another—anachronism as reality. To borrow a phrase from Jacques Derrida's *Specters of Marx*, this is "time out of joint."[35]

Economic growth is not the ultimate goal here, nor is dependence on the state. The fact that this sort of communal society existed in the past rhymes with the fact that it also arose as a possibility in the present. This is what Marx meant when he wrote in a letter to Engels dated March 25, 1868, that he'd found "what is newest in what is oldest."[36]

Marx detected a continuity between "old" communism and new communism. These were forms of communism completely different from that practiced by the Soviet Union and

China, which centers the state and government officials in its pursuit of modernization.

Marx's thought following the advent of the Paris Commune, which sought to abolish capital while delegitimizing state control, might best be termed "anarchist communism." (This term was likely first used by Joseph Déjacque, who was a fierce critic of Pierre-Joseph Proudhon—also a target of Marx's criticism—while heavily influenced by Peter Kropotkin, a Russian anarchist who wished to bring about a future society defined by degrowth.)

Despite what the word "anarchist" may suggest, this was not a form of radical individualism, nor was it a state of total lawlessness and disorder. Rather, anarchist communism was a form of "association-ism," in which social organization would occur from the bottom up, avoiding the hierarchical relations and domination demanded by capital and the state.

During the twentieth century, the attempt to rein in the power of capital took two primary forms: in the East, there was Soviet-style socialism, while in the West, there was the welfare state. They emerged in tandem and in competition with each other. As a result, these two methods converged their models of growth into one that was anchored in raising productivity through the centralized state control of the economy.

As alternatives to capitalism became restricted to frameworks that privileged growth, the emphasis on solidifying state power took center stage, separating Marxism from anarchism more and more. In this process we lost sight of how essential the apparatus of the commons was to Marx's later theorizing.

The Soviet Union no longer exists, and the welfare state is on the wane. When we look at "juridical illusions" like basic income and Modern Monetary Theory, it's clear that we have yet to break the spell of twentieth-century nationalism on our thinking, evidence of how impoverished our ability to imagine post-capitalism has become. This is why "anarchist communism" is so necessary.

THE POTENTIAL OF WORKERS' CO-OPS

In one of the few passages in *The Civil War in France* that includes the word "communism," Marx explains the kind of future society he saw become possible in the Paris Commune:

> If co-operative production is not to remain a sham and a snare; if it is to supersede the Capitalist system; if united co-operative societies are to regulate national production upon a common plan, thus taking it under their own control, and putting an end to the constant anarchy and periodical convulsions which are the fatality of Capitalist production—what else, gentlemen, would it be but Communism, "possible" Communism?[37]

Marx asserts that cooperative production is the basis of communism. It's clear that what he has in mind is entirely different from the Soviet-style state-centered "socialism."

But what does Marx mean by cooperative? While farmers' co-ops and trust unions are indeed cooperatives, what Marx is thinking of here are *workers' co-ops*.

A workers' co-op is comprised of members who contribute to and manage its governance communally and in doing so are able to control production through active participation in democratic decision-making processes. By abolishing wage labor conducted through employment and payment by capitalists, the co-op can instead be managed through the independent, democratic participation of the workers.

Workers' co-ops prioritize the production of essential items rather than superfluous luxury goods. They avoid excessive advertisement that attempts to add value solely through specious "novelty." Workers' co-ops don't make products like social media platforms and smartphone games that steal our time and data and make us addicted to our screens, and they emphasize using labor to meet the needs of the community. They make the knowledge and means necessary to production, as well as the products themselves, into integral parts of the commons.

It's important to view the restoration of the commons as more than a narrow, micro-level change. While workers' co-ops in Japan mainly focus on regional businesses like daycare, nursing, cleaning services, forestry, and the like, possible applications are endless. In places outside of Japan like Spain and Italy, this form of production has spread to industries as diverse as household appliance manufacture, agriculture, and publishing.

Workers' co-ops can partner with municipalities and cause-driven "social" businesses to construct broader, more expansive forms of association. In this vision, those managing various parts of the commons will act as a federation, helping each other adjust production to fit local needs.

Expanding communal production will allow for the displacement of capitalist production, which creates goods without any thought as to how much or even whether society needs them. Marx referred to the production of commodities based on the uncontrolled proliferation of divisions of labor peculiar to capitalism as "private labor." The purpose of a society founded on association is to eliminate private labor and eventually abolish wage labor entirely.

This might seem impossible to do in today's society with its complex division of labor. But with advances in information and communication technologies and the algorithms associated with them, such a transformation is no longer a distant possibility.

The global supply chain that feeds major industry already secures the goods we need on a large scale without subjecting companies to the whims of the free market. What I'm proposing is the transition of this production system into one actively managed by the workers themselves.

Of course, there's no need to continue production at its current scale. If we eliminate the excessive or unnecessary production of commodities overall in order to better reflect a balance of supply and demand centered on local production and consumption, then we can ensure that new technologies will be put to their most effective use.

TOWARD THE TRANSITION TO A USE-VALUE ECONOMY

As we promote decommodification and expand the commons, labor unions and workers' co-ops will rein in private labor, allowing us to shift from being a society that prioritizes unlimited economic growth above all to one based on use-value that engenders the fulfillment of people's actual needs.

During the pandemic, those of us in the Global North learned the hard way that the profit-seeking logic of capitalism can lead to the underproduction of necessary goods when we most need them. We should seize the chance to transition to a self-sufficient economy based on use-value. It is a form of self-preservation in a time when economic growth is under constant threat from, on the one hand, the vagaries of war, inflation, and climate-related crises, and on the other, ever-greater security risks like food and energy shortfalls and supply-chain disruptions.

In a use-value economy that has abandoned mass production and consumption, excessive advertising, incessant model upgrades, and planned obsolescence would no longer be necessary to keep the economy afloat. Jobs that generate no use-value in themselves, like those in marketing, advertising, and consultancy, would eventually wither away (or, at the very least, they would no longer pay many times more than actually essential work).

The remaining labor could be performed through work sharing, thereby reducing the length of the workday and increasing general free time. Adoption of a work-share paradigm would also allow us to abolish the divide between wage work and domestic labor. The latter would no longer be foisted predominantly on women, either—rather, everyone would participate in a work rotation, taking turns doing a variety of jobs in and out of the home.

If workers were able to choose what their labor produces and how that production takes place, the division between conception and execution would close. Technology would no longer be a tool to surveil and dominate workers but rather a mechanism for workers to overcome differences in skill level and work freely on their own terms. In other words, rather than being used to create unnecessary commodities, technological advancements could be utilized to shorten the workday, equalize skill levels, and remove strict hierarchies in the workplace.

Some jobs will still require highly skilled labor—not every job can become part of a work-share rotation. But there would no longer be a need to lavish unnecessarily high salaries on those whose jobs require specialized skill and expertise to perform. What if we created rewards and incentives out of fewer work hours instead?

The natural culmination of this kind of transition is degrowth communism. Under degrowth communism, education, health care, transportation, and so on would be available free of charge, while goods like food, clothing, books, and the like would circulate via mutual gift exchange. Job training, daycare, and general childcare support would be available to all, allowing everyone to live up to their full potential. This would be the true fulfillment of the promise articulated in the *Communist Manifesto* to create a new society based on "association, in which the free development of each is the condition for the free development of all."[38]

NOT PRIVATIZATION, BUT "PRIVATE CITIZEN-IZATION"

The fundamentals of a communism founded on the commons can in fact be found today, in the midst of our capitalist society. Any sort of mutual aid performed without the expectation of reciprocation—for example, helping your friend move—is an example of these communist principles.

Parents who serve food not only to their own children but also to their children's friends when they come over can be seen as participating in the gift economy. After all, do we ask those children's parents for compensation afterward?

No—instead we anticipate that our children will eventually end up eating at their house, or that they may give us some hand-me-down clothes, or do any other number of favors for us.

The point is that most of us are practicing communists already. Even at work, don't you help your colleagues out from time to time? Capitalism can't function without a little communism to help things along.

These practices of gift giving and mutual aid don't need to be restricted to family and friends. We should be able to make many different parts of our life function outside of the scope of commodities and money and instead become relations mediated by the commons.

There are already movements seeking to expand the commons even amid the breakneck advance of capitalism. There are initiatives recruiting private citizens to invest in the local production and consumption of electricity in the form of "people's power stations"; apps act as platforms for a "sharing economy" based on exchanging skills and goods. I see these kinds of movements as the "private citizen-ization" of society, as opposed to the "privatization" of society promoted by neoliberalism.

The advancement of private citizen-ization doesn't stop the exchange of goods and services as commodities in a money economy, of course; the marketplace still exists. But marketplaces existed even in precapitalist societies—there's no need to abolish them entirely. The defining feature of capitalism is its need to turn *everything* into a commodity, which drives the endless exploitation of nature and humankind in the name of valorizing capital.

The marketplace as it would exist after the expansion of the commons and private citizen-zation would be very different than the one that exists now. Because mutual aid practices would eliminate dependence on money and commodities, the motivation to maximize profits at all costs would reduce dramatically.

A SOCIETY OVERFLOWING WITH WEALTH

Marx discussed the richness that could be returned to our lives with the restoration of collective societal wealth. Let's go back and look at the paragraph that precedes the famous line previously quoted from *The Critique of the Gotha Program*.

> In a higher phase of communist society, after the enslaving subordination of the individual to the division of labor, and thereby also the antithesis between mental and physical labor, has vanished; after labor has become not only a means of life but life's prime want; after the productive forces have also increased with the all-round development of the individual, and all the springs of common wealth flow more abundantly—only then can the narrow horizon of bourgeois right be crossed in its entirety and society

> inscribe on its banners: From each according to his abilities, to each according to his needs![39]

In this passage Marx argues that communism will ultimately abolish the strict division of labor, and thereby the separation of conception and execution. Mass production driven by recklessly intensifying productivity in the name of profit seeking will also be left behind.

As the privatization, monopolization, and waste produced by capitalism wanes, the "springs of common wealth" will "flow more abundantly" for the common person. The richness of this societal wealth isn't measured by the sheer number of objects churned out by accelerating productivity levels. This is a conception of wealth still circumscribed by "the narrow horizon of bourgeois right." Rather, Marx is speaking of the abundance created when the commons is spread through work-sharing, mutual aid, and gift exchange—an abundance consisting of ample and equitable access to all human necessities.

This is not a type of wealth you can carelessly waste. Recall the lines quoted earlier from the third volume of *Capital* regarding the earth itself as a form of commons—we receive the earth from previous generations and have the obligation to pass it along to the generations to come. Natural resources and the planetary climate shouldn't be used to satisfy individual desires with no concern for others. The materiality of our "needs" arises out of relational respect for others and intentional mutual aid. The aim, then, is to be a society that allows everyone to become "fully developed individuals" (as discussed in chapter 3) with access to the forms of societal wealth most essential to them.

This transition would allow us to realize a society detached from the pursuit of economic growth—a degrowth society. For the first time, production would become truly sustainable, and the "irreparable rift" could, in fact, be repaired.

Today, mired in a twentieth-century conception of socialism, many critics dismiss the degrowth interpretation of Marxist thought. But there were others in Marx's time whose thought ran along similar lines.

Other socialists, affected by the Paris Commune, began to conceive of a future society that left consumerism behind and embraced collective prosperity. The English textile designer and socialist organizer William Morris was greatly influenced by Marx's writings and defined wealth as follows:

> Wealth is what Nature gives us and what a reasonable man can make out of the gifts of Nature for his reasonable use. The sunlight, the fresh air, the unspoiled face of the earth, food, raiment and housing necessary and decent; the storing up of knowledge of all kinds, and the power of disseminating it; means of free communication between man and man; works of art, the beauty which man creates when he is most a man, most aspiring and thoughtful—all things which serve the pleasure of people, free, manly, and uncorrupted. This is wealth.[40]

Morris critiqued the artificial scarcity created through capitalistic monopolies. He redefined wealth not as money, but as a form of "art" that will return joy to work and allow each person's talents to develop and thrive. Morris was active in labor union movements and worked to popularize this reconceptualization of the nature of "wealth."

The richness of wealth as Morris defines it resides in something completely separate from the wealth attainable through consumerism. It instead resonates with how Marx discussed wealth in the opening lines of *Capital*, and understandably so: It's said that Morris carried the French edition of *Capital* around with him everywhere until it grew tattered and worn.

For the nineteenth-century socialists who experienced the Paris Commune but had yet to see the Russian Revolution occur, a society based on association and the shared richness of communally conceived wealth seemed more realistic than the type of Soviet- or Chinese-style "socialism" we tend to think of today. This kind of expansive communist imagination is on display in Morris's own Paris Commune–inspired utopian novel, *News from Nowhere*.

WHY DID THE PARIS COMMUNE FALL?

Achieving the world described in *News from Nowhere* would be extremely challenging. After all, the Paris Commune itself existed for only a moment.

Isolated by police forces that had created a blockage around the Commune's perimeter, the Commune was extremely vulnerable to military suppression in May of 1871. Communication with the outside had to be conducted via balloons or homing pigeons, though eventually passports permitting travel between Paris and other regions were reinstated.

As a result, people in the provinces only got news about the social innovations taking place in the Commune via "the lies and slanders of the Versailles journals," as Marx puts it in his initial draft of *The Civil War in France*.[41] Even though the Commune sought to serve and aid farmers as well as city dwellers through the abolition of heavy taxation, mortgage debt, and the military draft, the Commune's political interests were described as being in opposition to those living in the provinces. As a result, farmers refrained from joining the Communards in their fight for independence and recognition.

It was clear that the Parisians alone would be unable to defeat the government army coming from Versailles. Unless city-dwelling workers could respond to the "raw interests" and "real needs" of those in the provinces and unite with them, they had very little chance of prevailing, and in the end, the conflict between the city and the country wasn't overcome.

In this respect, the Parisian situation resonates with the Russian one. The Russian agricultural communes known as the *mir* existed in scattered, isolated pockets. Marx understood that they could not gain the upper hand against the tsarist government by themselves.

In the first draft of his letter to Vera Zasulich, Marx describes the Russian *mir*'s status as a "localized microcosm" with a "lack of connection between the lives of different communes" as being a "debilitating feature" of the commune.[42] This is why Marx argued it was critical that Russia invest in the most essential innovations of western European capitalism such as railroads and telegraphs.

Overcoming the urban-rural divide thus became a fundamental concern for Marx later in his life. The urban-rural divide is not something that happens solely within a given state—it occurs at the local level (i.e., city vs. countryside), but also internationally, as the Global North versus the Global South. If we fail to address the issue at both scales, we'll never be able to truly fight capitalism.

Marx's research on the conflict continued throughout his later years. It's also what stood in the way of his completion of *Capital*. In the end, he failed to fully elucidate the communist society he envisioned, nor how it would truly materialize. But does that mean Marx's theorizing amounts to nothing more than an unrealizable fantasy?

ASSOCIATIONS ARE SPRINGING UP ALL OVER

The fact is, though, today there are many movements to form associations that can overcome precisely this sort of urban-rural divide. Global crises like the COVID-19 pandemic and climate change have enabled these movements to expand even further, both geographically and politically.

One such movement is "municipalism"—an international network of cities that originated in Barcelona. The basic unit of municipalism is the municipality as a unit of local government, and it seeks to expand political participation beyond just elections to include autonomous democracy and local consensus building.[43]

Amsterdam, one of the cities in this network, made waves by announcing that it plans to eliminate carbon emissions by 2050 and will adopt the "doughnut economics" model proposed by Oxford University economist Kate Raworth.

Doughnut economics is a framework for population well-being and environmental sustainability. The inner ring of the doughnut is the social foundation. If access to things like education, democracy, housing, electricity, and so on falls below a certain point, you end up in the void of human deprivation at the doughnut's center. The outer ring of the doughnut is the planetary boundary. If we overuse energy and natural resources, we'll exceed the limits of the planet's capacity to sustain human life. According to Raworth, it's essential that everyone live between these outer and inner rings—that is, within the doughnut itself.

But as wealth disparities widen under capitalism, many who live in the Global South are unable to reach the social foundation marked by the inner ring, while those pursuing boundless economic growth in the Global North enjoy decadent, resource-consuming lifestyles that break through the planetary boundary marked by the outer ring, destroying the planet in the process. Doughnut economics implies a transition to a degrowth economy.

Raworth herself doesn't advocate for communism. But doughnut economics nonetheless pushes us to fundamentally change how we think about the resources we have left.

For example, we tend to look at rising real estate prices from an investor's perspective, welcoming them as evidence of a city's successful development. But if we look at how rising prices threaten people's ability to find housing—in Amsterdam, for example, one out of every five people doesn't make enough to pay rent—we see that the calls for increased public housing, regulation of buying and selling real estate, and heavier taxation on investments are increasingly urgent concerns.

Furthermore, it is essential that we stop the exploitation of the Global South by the Global North. The mayor of Amsterdam, Femke Halsema, has openly called for the end of the exploitation and environmental destruction caused by the trade in West African cocoa, the clothing industry in Bangladesh, and the mining in Congo of the mineral coltan, which is used in laptops and smartphones.

How can this be done? Workers must be guaranteed stable employment even as we make the changes necessary to temper the inequality and economic precarity that leads them to buy cheap goods in the first place. In this way, the people of Amsterdam are working to conceive of an entirely new future for their city in order to simultaneously address the overlapping crises of environmental destruction and poverty.

There are many other initiatives in this vein. Influenced by municipalism, Berlin held a local referendum in September of 2021 that addressed the dramatic rent increases the city was suffering from. It was a bold initiative meant to require real estate companies that own buildings with three hundred or more apartments to sell a portion of the units to the Berlin government to then make them into public housing. A majority voted in favor for the proposal. Unfortunately the referendum wasn't legally binding, but the municipal and state government of Berlin is now obligated to investigate the proposal's feasibility. While the initiative remains in limbo, the fact that a public referendum endorsed such a radical proposal indicates that people are growing weary of capitalist principals.

These municipalities are reinventing themselves as autonomous bodies that are implementing reforms meant to benefit all citizens. They're fearlessly rejecting the commodification of societal wealth in the name of global capitalism, and the imposition of neoliberalism by national governments in the form of privatization and austerity measures. It might be too much for a single municipality to do alone. But as places like Amsterdam and Berlin attempt to stand up to global capitalism, they are creating links across both national boundaries and the divide between the Global North and South.

NOW IS THE TIME TO LEARN FROM MARX

Local communities and regional bodies are beginning to join forces to fight global capitalism—what does it all really mean, in the end?

As we endure economic inequality and climate change, pandemics and wars, more and more people, especially younger generations, feel that capitalism may well be reaching its conclusion. The trickle-down fairy tale that told us to endlessly expand the economy and heavily invest in technical developments in order to generate collective prosperity is losing its persuasiveness.

So, what sort of society—what sort of world—do we want to live in? And what are we willing to do to create it?

We continue to see citizen-centered movements springing up all over the world, creating international coalitions to expand the commons and protect our planet. Now is the time to learn from these activists, organizers, academics, and everyday people so we can all begin to imagine possibilities for the future with an unbiased, open mind.

Marx writes that the true driver of change in this world is the proletariat. All of us who feel burdened by the influence of capitalism are part of the proletariat.

Essential workers exhausted by long work hours and low pay, temporary workers stressed by the instability of their employment, salaried workers who've lost all sense of purpose in their jobs—not just at work but in our everyday lives at home we find ourselves subjected to the competitive principles and logic of the marketplace. Under capitalism, we are all part of the proletariat.

We feel capitalism's influence in racial and gender-based discrimination, in environmental crises and immigration-related issues. Of course, these problems would remain in the absence of capitalism. But capitalism by its nature produces inequalities that congeal into classes, intensifying existing problems like immiseration and poverty.

Capitalism generates endless disparities and division. The marketplace is designed to exclude and exploit the poor. This is why Marx advocates that we weaken the power of commodification and instead develop economic practices that are rooted in a participatory democracy. This is what communism is fighting for: a transition from the commodification of anything and everything to a *commonification* of all things.

MARX WAS A UTOPIAN THINKER

This book has invited us to reread Marx's *Capital* in a new light by studying new materials that elaborated on how his thinking evolved near the end of his life. Perhaps, along the way, your perception of Marx has changed, too.

Marx does not hold a magic key that will solve all our economic and social problems. But he is a systematic thinker whose work can help us choose a less destructive and more equitable path to a better world.

It's my sincerest hope that reading *Capital* in a time of crisis like our own can be a catalyst for overcoming the "common sense" of capitalism and instead restoring our capacity to imagine a richer lifestyle, all while inviting us to take necessary steps to make this imagined future a reality.

Why is the nineteenth-century thinking of someone like Marx still relevant now? Marx was a revolutionary and utopian thinker. Much of postwar thinking assumed that without the instability of revolution or war, economic growth could continue exponentially. This sort of optimism popularized reformist innovations like the welfare state.

But such optimism about the future no longer seems justified. There are still those who tirelessly assert their belief in humanity's capacity for continuous progress, but in the face of the many overlapping crises that have come to define our time, a future of chaos and confusion seems inevitable.

The fact is, the age of neoliberalism is over. Times like ours call for a stronger state. But if emergencies like climate catastrophe, deepening inequality, and dangerous wars and pandemics continue to proliferate, the state will only grow more powerful until it descends into full-blown fascism or totalitarianism. We cannot allow the reemergence of political figures like Hitler or Stalin.

To avoid this sort of barbarism, we must confront the defining problems of our age—inequality and exploitation, war and violence, colonization and slavery—by learning from thinkers like Marx, whose knowledge and powers of imagination will give us the tools we need to fight the accelerating incursions of state power and create a more just and free world.

AFTERWORD

In a Time of Revolution

Let's look back at the 2010s. How do you remember them? To me, they seem like a decade of revolution.

Revolution? Where? Isn't the word itself a bit outdated?

In Japan, the beginning of the 2010s saw a broad-based antinuclear movement rise up in the wake of the Fukushima triple disaster of 2011. Huge demonstrations rocked a nation that was widely said to have abandoned protesting. For me, personally, this marked the beginning of my interest in the intersection of Marxism and the natural world.

Yet nuclear plants continue to operate. Due to strains on the power supply, there are calls not only for the revival of decommissioned plants but even the construction of new ones. The antinuclear struggle has ended in defeat.

Around the same time as the post-Fukushima antinuclear protests, the United States saw the emergence of the Occupy Wall Street movement. Under the slogan *We Are the 99 Percent*, young people occupied Zuccotti Park in New York City to decry the hoarding of wealth by the richest 1 percent of the population.

In many respects this movement ended in defeat as well, broken up by the police after only a few months. Yet who would have imagined that a movement so harshly critical of capitalism would gather so much support in the United States?

The inspiration for Occupy Wall Street was the Arab Spring of 2011. Triggered by the self-immolation of a young man in Tunisia who was dissatisfied with the economic inequality in his country, uprisings broke out across the Middle East and Africa, toppling decades-old dictatorial regimes such as the Mubarak government in Egypt and the Gaddafi regime in Libya. However, even here, with the exception of Tunisia's "Jasmine Revolution," these revolutions were defeated and dictatorships restored.

Protests spread to Europe, too. In Spain, scores of *indignados* joined the anti-austerity 15-M movement, which led to the election of the left-wing Podemos party. During the 2012 Greek government-debt crisis, the left-wing coalition Syriza came to power and challenged the "troika" of the IMF (International Monetary Fund), EC (European Commission), and ECB (European Central Bank), all of which had instituted drastic austerity measures in an effort to resolve the country's financial struggles (though this, too, ended in defeat).

Pro-democracy movements emerged in Asia, such as the Sunflower Movement in Taiwan, the Umbrella Movement in

Hong Kong, and the Milk Tea Alliance in Thailand and Myanmar.

Major anti-government protests emerged in Iran in response to the murder of Mahsa Amini, a twenty-two-year-old woman who was arrested by the morality police for "inadequately" covering her hair with her hijab. Women are risking their lives in a brutal fight for freedom and equality. While the movement was largely suppressed by the government, it still points to the possibility of a new, women-led revolution, the first of its kind in world history.

Since the collapse of the Soviet Union in 1991, global capitalism has led to a considerable decline of leftist social movements. However, starting in the 2010s, movements like the ones just described, which disavowed capitalism, white supremacy, war, patriarchy, environmental destruction, inequality, fascism and more, have emerged all over the world.

Socialism, too, has become an increasingly popular proposition, especially among millennials and Gen Z. Look no further than the self-identified thirty-four-year-old socialist Zohran Mamdani, whose landslide victory in the New York City mayoral race marked an unprecedented embrace of socialist principles and policies. His platform advocates for expanding the commons in the form of free buses, free childcare, and lower rents.

I welcome this trend, but I remain concerned. I was in my twenties during the 2008 economic crisis, which inspired my decision to study Marx and deepen my understanding of capitalism. I've been greatly inspired and encouraged by these revolutionary attempts to envision a new society. Even as they suffer defeat, these are comrades striving to realize a just society—sometimes at the risk of their lives. I stand in solidarity with them.

But there's no guarantee that popular dissatisfaction itself will lead to the creation of a better society. The 2010s were both a revolutionary time and a reactionary one. All over the world, there are oppressive forces bearing down on calls for democracy and justice.

It is essential that we do not turn on one other and allow differences in race, nationality, and gender to divide the proletariat. We must demonstrate convincingly that there are better paths to our collective prosperity. This is no easy task: It's very difficult to question what is generally taken for granted, to conceive of alternatives and act accordingly. Even so, someone needs to take the first step.

This book is an attempt to use Marx's *Capital* to take such a step. This book's interpretation is by no means the last word on *Capital*. Interpretations of Marxism will inevitably vary based on the concerns of the reader and their context. That's what makes a classic work like this so fascinating: There is a range of perspectives on his work and theorizing, and it's important to consider them all as we navigate an ever-changing world.

We are living through a moment when the old answers no longer hold, and the future is up for grabs. *Capital* will not offer a ready-made blueprint, but it can sharpen our sense of what is possible, and what must be transformed. I hope this book encourages you to approach Marx not as a monument but as a tool—one you can pick up, challenge, and use in your own struggles to build the world to come.

NOTES

1 All quotes from *Capital* are from: Karl Marx, *Capital*, vol. 1, Ben Fowkes, trans. (Penguin Classics, 1990). All subsequent references appear parenthetically within the text and refer to page numbers from this edition.

2 The *Rheinische Zeitung* ran from January of 1842 to March of 1843, and was published in the city of Cologne (Köln) in Prussia (modern-day Germany). The newspaper was founded when a faction of cutting-edge young Hegelian intellectuals like Moses Hess (these were leftist Hegelians—Marx was one of them) convinced a group of wealthy liberal capitalists to invest in it. Marx became the editor in chief of the newspaper in November 1842, and subsequently increased its circulation by nearly four times. However, the Prussian government could not tolerate the level of criticism directed at the administration and state, and it forced publication to cease in March of 1843.

3 Michael Ende, *Momo*, J. Maxwell Brownjohn, trans. (Penguin, 1984), 58.

4 Yoshihiko Uchida, *Shihonron no sekai* (*The World of* Capital), (Iwanami Shoten, 1966), 78. Translation of excerpt by Brian Bergstrom.

5 Makoto Kumazawa, *Minshushugi wa kōjō no monzen de tachisukumu* (*Democracy Stops at the Factory Gate*), (Shakai shisō-sha, 1993).

6 Karl Marx, "Results of the Direct Production Process," *Marx and Engels Collected Works*, vol. 34: *Marx 1861–64*, Ben Fowkes, trans. (Lawrence & Wishart, 2010), 435. Marx originally planned to include "Results of the Direct Production Process" as a section of *Capital*, but it did not end up in the final version.

7 John Maynard Keynes, "Economic Possibilities for Our Grandchildren," *Essays in Persuasion* (Palgrave Macmillan, 2010 [o.p.d.: 1931]), 321–32.

8 Keynes, "Economic Possibilities," 322.

9 Karl Marx, "Human Requirements and Division of Labor Under the Rule of Private Property," *Economic and Philosophic Manuscripts of 1844*, Martin Milligan, trans. (Progress Publishers, 1959), 110. This manuscript was originally written in 1844 but first published in 1932.

10 Harry Braverman, *Labor and Monopoly Capital: The Degradation of Work in the Twentieth Century* (Monthly Review Press, 1974).

11 David Graeber, *Bullshit Jobs: A Theory* (Simon & Schuster, 2019).

12 Tatsushi Fujihara, *Kyūshoku no rekishi* (*The History of School Lunch*), (Iwanami Shoten, 2018).

13 Karl Marx, *Marx-Engels-Gesamtausgabe II, Abteilung Band 4.2* (Dietz Verlag, 1993), 752–3. This passage appears slightly differently in the presently published version of *Capital*, vol. 3, David Fernbach, trans. (Penguin Classics, 1991), on page 949:

> On the other hand, large landed property reduces the agricultural population to an ever decreasing minimum and confronts it with an ever growing industrial population crammed together in large towns; in this way, it produces conditions that provoke an irreparable rift in the interdependent process of social metabolism, a metabolism prescribed by the natural laws of life itself. The result of this is a squandering of the vitality of the soil, which is carried by trade far beyond the bounds of a single country (Liebig).

The parenthetical citation of Liebig in both cases is in the original.

14 Karl Marx, *A Contribution to the Critique of Political Economy*, S. W. Ryazanskaya, trans. (Progress Publishers, 1977), 21.

15 Karl Marx, "Letter from Marx to His Father, November 1837," *The First Writings of Karl Marx*, Paul M. Schafer, ed. and trans. (Ig Publishing, 2006), 77.

16 For more on this, please see Kōhei Saitō, *Karl Marx's Ecosocialism: Capital, Nature, and the Unfinished Critique of Political Economy* (Monthly Review Press, 2017).

17 Karl Marx, "Letter from Marx to Engels, March 25, 1868," John Peet, trans., *MECW* vol. 42: *Letters 1864–68* (Lawrence & Wishart, 2010), 559.

18 Marx, *Capital* vol. 3, 959.

19 Makoto Nishibe, *Whither Capitalism?: Internalizing the Market and Free Investment* (Springer, 2019).

20 Kōhei Saitō and Takeshi Yoro, "Yōrō Takeshi Saitō Kōhei taidan: 'taru o shiru' ikikata ga sekai o sukuu" ("Knowing When Enough is Enough: How the Way We Live Can Save the World, a Conversation between Kōhei Saitō and Takeshi Yōrō"), October 16, 2021, https://jbpress.ismedia.jp/articles/-/67256?page=7.

21 Fredric Jameson, "An American Utopia," in Slavoj Žižek, ed., *An American Utopia: Dual Power and the Universal Army* (Verso, 2016), 3.

22 Ben Westcott and Yong Xiong, "Marxist Student Snatched on Way to Mao Zedong Celebration in China," CNN World, December 26, 2018 (accessed online September 9, 2025).

23 See, for example, Teinosuke Ōtani, ed., *Soren no shakaishugi wa nan datta no ka* (*What Exactly Was Soviet Socialism?*) (Ōtsuki shoten, 1996).

24 Immanuel Wallerstein, "Marxisms," in Marcello Musto, ed., *The Marx Revival* (Cambridge University Press, 2020), 376–89.

25 Branko Milanovic, *Capitalism, Alone* (Harvard University Press, 2019).

26 See, for example, Gøsta Esping-Andersen, *The Three Worlds of Welfare Capitalism* (Princeton University Press, 1990).

27 Thomas Piketty, *Capital in the Twenty-First Century*, Arthur Goldhammer, trans. (Harvard University Press, 2014).

28 See, for example, David Marchese, "Thomas Piketty Thinks America Is Primed for Wealth Redistribution," *The New York Times*, April 1, 2022.

29 Ulrich Brand and Markus Wissen, *Imperiale Lebensweise: Zur Ausbeutung von Mensch und Natur im Globalen Kapitalismus* (oekom, 2017), 64–5. See also: Kōhei Saitō, *Slow Down: The Degrowth Manifesto*, Brian Bergstrom, trans. (Astra House, 2024), 8–10.

30 Robert L. Heilbroner, *The Making of Economic Society* (Prentice Hall, 1993).

31 Karl Marx, "Drafts of the Letter to Vera Zasulich: First Draft," Barrie Selman, trans., *MECW* vol. 24: *Marx and Engels: 1874–1883* (Lawrence & Wishart, 2010), 352–3, 354.

32 Karl Marx, "Drafts of a Reply: The 'First' Draft," Teodor Shanin, trans., *Late Marx and the Russian Road: Marx and "the Peripheries of Capitalism"* (Monthly Review Press, 1983), 106–7. See also, for an alternative translation, *MECW* vol. 24, 349–50.

33 In May of 1875, when the General German Workingmen's Association (Lassalle faction) and the Social Democratic Workers' Party (the Eisenachers) held a joint conference in the city of Gotha to form the German Socialist Workers' Party, its manifesto was called the Gotha Program. Marx, sensing the growing influence of his rival Ferdinand Lassalle, published this treatise as a "critique" of this program.

34 Karl Marx and Friedrich Engels, "Preface to the German Edition of 1872," *Manifesto of the Communist Party* (Progress Publishers, 1977), 12.

35 Jacques Derrida, *Specters of Marx: The State of the Debt, the Work of Mourning and the New International*, Peggy Kamuf, trans. (Routledge, 1994).

36 Marx, "Letter from Marx to Engels, March 25, 1868," 557.

37 Karl Marx, "The Civil War in France," *MECW* vol. 22: *Marx and Engels: 1870–71* (Lawrence & Wishart, 2010), 335.

38 Karl Marx and Friedrich Engels, *Manifesto of the Communist Party* (Progress Publishers, 1977), 60.

39 Karl Marx, "The Critique of the Gotha Programme," Peter and Betty Ross, trans., *MECW* vol. 24: *Marx and Engels 1874–83* (Lawrence & Wishart, 2010), 87.

40 William Morris, *Useful Work versus Useless Toil* (Hammersmith Socialist Society, 1893), 7.

41 Karl Marx, "First Draft of *The Civil War in France*," *MECW* vol. 22: *Marx and Engels: 1870–71* (Lawrence & Wishart, 2010), 466.

42 Marx, "Drafts of a Reply," 111.

43 For a more detailed description and analysis of the municipalism movement, see Saitō, *Slow Down*, 212–21.

INDEX

ABOUT THE AUTHOR

Kōhei Saitō is an associate professor of philosophy at the University of Tokyo. He received his PhD in philosophy from Humboldt-Universität zu Berlin in 2016. He was awarded the 2018 Isaac Deutscher Memorial Prize, the most prestigious academic award for Marxian studies, making Saitō its youngest recipient. In 2020, the Japan Society for the Promotion of Science awarded him the highly prestigious JSPS prize, awarded to the top twenty-five scholars (only a few in humanities and social sciences) in the country under the age of forty-five. *Slow Down: The Degrowth Manifesto* (originally published in Japanese as *Capital in the Anthropocene*) received the Best Asian Books of the Year prize from the Asia Book Awards in 2021.

ABOUT THE TRANSLATOR

Brian Bergstrom is a lecturer and translator who has lived in Chicago, Kyoto, and Yokohama. His writing and translations have appeared in publications including *Granta*, *The Paris Review*, *Aperture*, *Lit Hub*, and *The Penguin Book of Japanese Short Stories.* His translation of *Trinity, Trinity, Trinity* by Erika Kobayashi won the 2022 JUSFC Prize for the Translation of Japanese Literature, and his translation of *Slow Down: The Degrowth Manifesto* by Kōhei Saitō was short-listed for the 2024 Sasakawa Foundation Translation Prize. His newest book-length translation is *The Dilemmas of Working Women* by Fumio Yamamoto (HarperVia, 2025). He is currently based in Montréal, Canada.